50% OFF Online ASVAB Course.

Dear Customer,

We consider it an honor and a privilege that you chose our ASVAB Study Guide. As a way of showing our appreciation and to help us better serve you, we have partnered with Mometrix Test Preparation to offer you **50% off their online ASVAB Prep Course.** Many ASVAB courses are needlessly expensive and don't deliver enough value. With their course, you get access to the best ASVAB prep material, and you only pay half price.

Mometrix has structured their online course to perfectly complement your printed study guide. The ASVAB Online Course contains **in-depth lessons** that cover all the most important topics, **210+ video reviews** that explain difficult concepts, over **1,100 practice questions** to ensure you feel prepared, and **300+ digital flashcards**, so you can study while you're on the go.

Online ASVAB Prep Course

Topics Include:

- General Science
- Arithmetic Reasoning
- Mathematics Knowledge
- Word Knowledge
- Paragraph Comprehension
- Electronics Information
- Auto Information
- Shop Information
- Mechanical Comprehension
- Assembling Objects

Course Features:

- ASVAB Study Guide
  - Get content that complements our best-selling study guide.
- Full-Length Practice Tests
  - With over 1100 practice questions, you can test yourself again and again.
- Mobile Friendly
  - If you need to study on the go, the course is easily accessible from your mobile device.
- ASVAB Flashcards
  - Their course includes a flashcards mode with over 300 content cards for you to study.

To receive this discount, visit their website: mometrix.com/university/asvab and add the course to your cart. You can also scan the QR code with your smart phone. At the checkout page, enter the discount code: **APEXASVAB50**

If you have any questions or concerns, please contact them at universityhelp@mometrix.com.

Sincerely,

 in partnership with **Mometrix** TEST PREPARATION

FREE

Free Study Tips Videos/DVD

In addition to this guide, we have created a FREE set of videos with helpful study tips. **These FREE videos provide you with top-notch tips to conquer your exam and reach your goals.**

Our simple request is that you give us feedback about the book in exchange for these strategy-packed videos. We would love to hear what you thought about the book, whether positive, negative, or neutral. It is our #1 goal to provide you with quality products and customer service.

To receive your **FREE Study Tips Videos**, scan the QR code or email freevideos@apexprep.com. Please put "FREE Videos" in the subject line and include the following in the email:

    a. The title of the book

    b. Your rating of the book on a scale of 1-5, with 5 being the highest score

    c. Any thoughts or feedback about the book

Thank you!

ASVAB Study Guide 2023-2024

3 ASVAB Practice Tests and Exam Prep Book for All Military
Branches (Marines, Navy, Army, Air Force, Coast Guard)
[3rd Edition]

J. M. Lefort

Written and edited by APEX Publishing.

ISBN 13: 9781637756577
ISBN 10: 1637756577

APEX Publishing is not connected with or endorsed by any official testing organization. APEX Publishing creates and publishes unofficial educational products. All test and organization names are trademarks of their respective owners.

The material in this publication is included for utilitarian purposes only and does not constitute an endorsement by APEX Publishing of any particular point of view.

For additional information or for bulk orders, contact info@apexprep.com.

Table of Contents

Electronics Information 172
Practice Quiz 187
Answer Explanations 188

Auto Information 190
Practice Quiz 202
Answer Explanations 204

Shop Information 206
Types of common shop tools and their uses...206
Shop PPE and Safety 208
Practice Quiz 210
Answer Explanations 211

Mechanical Comprehension 212
Practice Quiz 222
Answer Explanations 223

Assembling Objects 225
Practice Quiz 227
Answer Explanations 229

Practice Test #1 230
General Science 230
Arithmetic Reasoning 232
Word Knowledge 234
Paragraph Comprehension 236
Mathematics Knowledge 240
Electronics Information 243
Auto Information 247
Shop Information 248
Mechanical Comprehension 250
Assembling Objects 253

Answer Explanations #1 259
General Science 259
Arithmetic Reasoning 260
Word Knowledge 262
Paragraph Comprehension 263
Mathematics Knowledge 264

Welcome

Dear Customer

Congratulations on taking the next step in your educational journey, and thank you for choosing APEX to help you prepare! We are delighted to be by your side, equipping you with the knowledge and skills needed to make this move forward. Your APEX study guide contains helpful tips and quality study material that will contribute to your success. This study guide has been tailored to assist you in passing your chosen exam, but it also includes strategies to conquer any test with ease. Whether your goal is personal growth or acing that big exam to move up in your career, our goal is to leave you with the confidence and ability to reach the top!

We love to hear success stories, so please let us know how you do on your exam. Since we are continually making improvements to our products, we welcome feedback of any sort. Your achievements as well as criticisms can be emailed to info@apexprep.com.

Sincerely,
APEX Team

FREE Videos/DVD OFFER

Achieving a high score on your exam depends on both understanding the content and applying your knowledge. **Because your success is our primary goal, we offer FREE Study Tips Videos, which provide top-notch test taking strategies to help optimize your testing experience.**

Our simple request is that you email us feedback about our book in exchange for the strategy packed videos.

To receive your **FREE Study Tips Videos**, scan the QR code or email freevideos@apexprep.com. Please put "FREE Videos" in the subject line and include the following in the email:

    a. The title of the book
    b. Your rating of the book on a scale of 1-5, with 5 being the highest score
    c. Any thoughts or feedback about the book

Thank you!

Test Taking Strategies

1. Reading the Whole Question

A popular assumption in Western culture is the idea that we don't have enough time for anything. We speed while driving to work, we want to read an assignment for class as quickly as possible, or we want the line in the supermarket to dwindle faster. However, speeding through such events robs us from being able to thoroughly appreciate and understand what's happening around us. While taking a timed test, the feeling one might have while reading a question is to find the correct answer as quickly as possible. Although pace is important, don't let it deter you from reading the whole question. Test writers know how to subtly change a test question toward the end in various ways, such as adding a negative or changing focus. If the question has a passage, carefully read the whole passage as well before moving on to the questions. This will help you process the information in the passage rather than worrying about the questions you've just read and where to find them. A thorough understanding of the passage or question is an important way for test takers to be able to succeed on an exam.

2. Examining Every Answer Choice

Let's say we're at the market buying apples. The first apple we see on top of the heap may *look* like the best apple, but if we turn it over we can see bruising on the skin. We must examine several apples before deciding which apple is the best. Finding the correct answer choice is like finding the best apple. Although it's tempting to choose an answer that seems correct at first without reading the others, it's important to read each answer choice thoroughly before making a final decision on the answer. The aim of a test writer might be to get as close as possible to the correct answer, so watch out for subtle words that may indicate an answer is incorrect. Once the correct answer choice is selected, read the question again and the answer in response to make sure all your bases are covered.

3. Eliminating Wrong Answer Choices

Sometimes we become paralyzed when we are confronted with too many choices. Which frozen yogurt flavor is the tastiest? Which pair of shoes look the best with this outfit? What type of car will fill my needs as a consumer? If you are unsure of which answer would be the best to choose, it may help to use process of elimination. We use "filtering" all the time on sites such as eBay® or Craigslist® to eliminate the ads that are not right for us. We can do the same thing on an exam. Process of elimination is crossing out the answer choices we know for sure are wrong and leaving the ones that might be correct. It may help to cover up the incorrect answer choice. Covering incorrect choices is a psychological act that alleviates stress due to the brain being exposed to a smaller amount of information. Choosing between two answer choices is much easier than choosing between all of them, and you have a better chance of selecting the correct answer if you have less to focus on.

4. Sticking to the World of the Question

When we are attempting to answer questions, our minds will often wander away from the question and what it is asking. We begin to see answer choices that are true in the real world instead of true in the world of the question. It may be helpful to think of each test question as its own little world. This world may be different from ours. This world may know as a truth that the chicken came before the egg or may assert that two plus two equals five. Remember that, no matter what hypothetical nonsense may be in the question, assume it to be true. If the question states that the chicken came before the egg, then choose

2

your answer based on that truth. Sticking to the world of the question means placing all of our biases and assumptions aside and relying on the question to guide us to the correct answer. If we are simply looking for answers that are correct based on our own judgment, then we may choose incorrectly. Remember an answer that is true does not necessarily answer the question.

5. Key Words

If you come across a complex test question that you have to read over and over again, try pulling out some key words from the question in order to understand what exactly it is asking. Key words may be words that surround the question, such as *main idea, analogous, parallel, resembles, structured,* or *defines.* The question may be asking for the main idea, or it may be asking you to define something. Deconstructing the sentence may also be helpful in making the question simpler before trying to answer it. This means taking the sentence apart and obtaining meaning in pieces, or separating the question from the foundation of the question. For example, let's look at this question:

> Given the author's description of the content of paleontology in the first paragraph, which of the following is most parallel to what it taught?

The question asks which one of the answers most *parallels* the following information: The *description* of paleontology in the first paragraph. The first step would be to see *how* paleontology is described in the first paragraph. Then, we would find an answer choice that parallels that description. The question seems complex at first, but after we deconstruct it, the answer becomes much more attainable.

6. Subtle Negatives

Negative words in question stems will be words such as *not, but, neither,* or *except.* Test writers often use these words in order to trick unsuspecting test takers into selecting the wrong answer—or, at least, to test their reading comprehension of the question. Many exams will feature the negative words in all caps (*which of the following is NOT an example*), but some questions will add the negative word seamlessly into the sentence. The following is an example of a subtle negative used in a question stem:

> According to the passage, which of the following is *not* considered to be an example of paleontology?

If we rush through the exam, we might skip that tiny word, *not,* inside the question, and choose an answer that is opposite of the correct choice. Again, it's important to read the question fully, and double check for any words that may negate the statement in any way.

7. Spotting the Hedges

The word "hedging" refers to language that remains vague or avoids absolute terminology. Absolute terminology consists of words like *always, never, all, every, just, only, none,* and *must.* Hedging refers to words like *seem, tend, might, most, some, sometimes, perhaps, possibly, probability,* and *often.* In some cases, we want to choose answer choices that use hedging and avoid answer choices that use absolute terminology. It's important to pay attention to what subject you are on and adjust your response accordingly.

8. Restating to Understand

Every now and then we come across questions that we don't understand. The language may be too complex, or the question is structured in a way that is meant to confuse the test taker. When you come across a question like this, it may be worth your time to rewrite or restate the question in your own words in order to understand it better. For example, let's look at the following complicated question:

> Which of the following words, if substituted for the word *parochial* in the first paragraph, would LEAST change the meaning of the sentence?

Let's restate the question in order to understand it better. We know that they want the word *parochial* replaced. We also know that this new word would "least" or "not" change the meaning of the sentence. Now let's try the sentence again:

> Which word could we replace with *parochial,* and it would not change the meaning?

Restating it this way, we see that the question is asking for a synonym. Now, let's restate the question so we can answer it better:

> Which word is a synonym for the word *parochial*?

Before we even look at the answer choices, we have a simpler, restated version of a complicated question.

9. Predicting the Answer

After you read the question, try predicting the answer *before* reading the answer choices. By formulating an answer in your mind, you will be less likely to be distracted by any wrong answer choices. Using predictions will also help you feel more confident in the answer choice you select. Once you've chosen your answer, go back and reread the question and answer choices to make sure you have the best fit. If you have no idea what the answer may be for a particular question, forego using this strategy.

10. Avoiding Patterns

One popular myth in grade school relating to standardized testing is that test writers will often put multiple-choice answers in patterns. A runoff example of this kind of thinking is that the most common answer choice is "C," with "B" following close behind. Or, some will advocate certain made-up word patterns that simply do not exist. Test writers do not arrange their correct answer choices in any kind of pattern; their choices are randomized. There may even be times where the correct answer choice will be the same letter for two or three questions in a row, but we have no way of knowing when or if this might happen. Instead of trying to figure out what choice the test writer probably set as being correct, focus on what the *best answer choice* would be out of the answers you are presented with. Use the tips above, general knowledge, and reading comprehension skills in order to best answer the question, rather than looking for patterns that do not exist.

Introduction

Function of the Test

The Armed Services Vocational Aptitude Battery (ASVAB) is an exam created by the Defense Department to test a person's potential regarding military training. Individuals may take the ASVAB as an entrance exam into the military (MEP), or they may take it as a career planning program (CEP) with no commitment to the military. For the ASVAB CEP, high school students in 10th, 11th, and 12th grade may participate, usually to find out what strengths they possess for future careers. The ASVAB MEP is for students who have at least graduated high school or have their GED. For students who have their GED, a higher score is required for entrance into the military.

Test Administration

There are two versions of the ASVAB: the Military Entrance Processing (MEP) and the Career Exploration Program (CEP). The MEP, given as an entrance exam, requires communication with a recruiter for scheduling purposes. The ASVAB MEP is taken on a computer, while the ASVAB CEP is taken on a computer or with pencil and paper at participating high schools. Individuals may retake the ASVAB exam, but they must wait one month to do so. After the second retake, another one-month wait is required. A third retake, and any subsequent retake, will require a six-month wait.

Test Format

The ASVAB computer-based test is an adaptive test that changes difficulty based on an individual's answers. The paper version of the ASVAB is similar, except the answers do not adapt to your skill level as you progress. This guide is based on the computer version of the exam. There are nine sections in the ASVAB.

Scoring

For ASVAB scoring, mean and standard deviation of the subtest scores are used to find the overall score. The subtests are on a range from one to 100, so the mean is set to 50 with a standard deviation set to 10. Each branch of the military requires a different minimum score. Here are some examples below:

- Air Force minimum score: 36
- Army minimum score: 31
- Coast Guard minimum score: 40
- Marine Corps minimum score: 32
- Navy minimum score: 35

Bonus Content

Practice Test #1 and Practice Test #2 can be found in this book and online. Practice Test #3 is exclusively online. To access the online practice tests along with other features, type in the link below or scan the QR code:

apexprep.com/bonus/asvab

The first time you access the page, you will need to register as a "new user" and verify your email address. If you encounter any problems, please email info@apexprep.com.

Study Prep Plan

Breathe

Reducing stress is key when preparing for your test.

Build

Create a study plan to help you stay on track.

Begin

Stick with your study plan. You've got this!

1 Week Study Plan

Day 1	Day 2	Day 3	Day 4	Day 5	Day 6	Day 7
General Science	Arithmetic Reasoning	Mathematics Knowledge	Electronics Information	Practice Tests #1 & #2	Practice Test #3	Take Your Exam!

2 Week Study Plan

Day 1	Day 2	Day 3	Day 4	Day 5	Day 6	Day 7
General Science	Arithmetic Reasoning	Word Knowledge	Mathematics Knowledge	Geometry and Measurement	Electronics Information	Shop Information

Day 8	Day 9	Day 10	Day 11	Day 12	Day 13	Day 14
Practice Test #1	Answer Explanations #1	Practice Test #2	Answer Explanations #2	Practice Test #3	Answer Explanations #3	Take Your Exam!

30 Day Study Plan

Day 1	Day 2	Day 3	Day 4	Day 5	Day 6	Day 7
General Science	Physics	Biology	Arithmetic Reasoning	Arithmetic Operations with Rational Numbers	Applying Estimation Strategies and Rounding Rules to Real-World Problems	Operations and Properties of Rational Numbers

Day 8	Day 9	Day 10	Day 11	Day 12	Day 13	Day 14
Word Knowledge	Paragraph Comprehension	Mathematics Knowledge	Solving a Linear Inequality in One Variable	Graphing and Statistics	Important Features of Graphs	Geometry and Measurement

Day 15	Day 16	Day 17	Day 18	Day 19	Day 20	Day 21
Area and Perimeter	Similarity, Right Triangles, and Trigonometric Ratios	Electronics Information	Parallel Circuits	Auto Information	Automotive Components	Shop Information

Day 22	Day 23	Day 24	Day 25	Day 26	Day 27	Day 28
Mechanical Comprehension	Assembling Objects	Practice Test #1	Answer Explanations #1	Practice Test #2	Answer Explanations #2	Practice Test #3

Day 29	Day 30
Answer Explanations #3	Take Your Exam!

General Science

Chemistry

Scientific Notation and Temperature Scales

Scientific Notation

Scientific notation is a system used to represent numbers that are very large or very small. Sometimes numbers are way too big or small to be written out with all their digits, so scientific notation is used to express these numbers in a simpler way.

Scientific notation takes a number's decimal notation and turns it into exponent form, as shown in the table below:

Decimal Notation	Scientific Notation
5	5×10^0
500	5×10^2
10,000,000	1×10^7
8,000,000,000	8×10^9
−55,000	-5.5×10^4
.00001	10^{-5}

In scientific notation, the decimal is placed after the first digit and all the remaining numbers are dropped. For example, 5 becomes "5.0 × 10^0." This equation is raised to the zero power because there are no zeros behind the number "5." Always put the decimal after the first number. Let's say you have the number 125,000. We would write this in scientific notation as 1.25×10^5. Multiplying by 10^5 essentially means moving the decimal point to the right five times. Therefore, writing in scientific notation allows you to write the number in three decimal places instead of five. The metric system's base units are meter for length, kilogram for mass, and liter for liquid volume.

Temperature Scales

Science utilizes three primary **temperature scales**. The temperature scale most often used in the United States is the **Fahrenheit (°F) scale**. The Fahrenheit scale uses key markers based on the measurements of the freezing (32 °F) and boiling (212 °F) points of water. In the United States, when taking a person's temperature with a thermometer, the Fahrenheit scale is used to represent this information. The human body registers an average temperature of 98.6 °F.

Another temperature scale commonly used in science is the **Celsius (°C) scale** (also called *centigrade* because the overall scale is divided into one hundred parts). The Celsius scale marks the temperature for water freezing at 0 °C and boiling at 100 °C. The average temperature of the human body registers at 37 °C. Most countries in the world use the Celsius scale for everyday temperature measurements.

For scientists to easily communicate information regarding temperature, an overall standard temperature scale was agreed upon. This scale is the **Kelvin (K) scale**. Named for Lord Kelvin, who conducted research in thermodynamics, the Kelvin scale contains the largest range of temperatures to facilitate any possible readings.

The Kelvin scale is the accepted measurement by the International System of Units (from the French *Système international d'unités*), or SI, for temperature. The Kelvin scale is employed in thermodynamics, and its reading for 0 is the basis for absolute zero. This scale is rarely used for measuring temperatures in the medical field.

The conversions between the temperature scales are as follows:

Degrees Fahrenheit to Degrees Celsius:

$$°C = \frac{5}{9}(°F - 32)$$

Degrees Celsius to Degrees Fahrenheit:

$$°F = \frac{9}{5}(°C) + 32$$

Degrees Celsius to Kelvin:

$$K = °C + 273.15$$

For example, if a patient has a temperature of 38 °C, what would this be on the Fahrenheit scale?

Solution:

First, select the correct conversion equation from the list above.

$$°F = \frac{9}{5}(°C) + 32$$

Next, plug in the known value for °C, 38.

$$°F = \frac{9}{5}(38) + 32$$

Finally, calculate the desired value for °F.

$$°F = \frac{9}{5}(38) + 32$$

$$°F = 100.4°F$$

For example, what would the temperature 52 °C be on the Kelvin scale?

First, select the correct conversion equation from the previous list.

$$K = °C + 273.15$$

Next, plug in the known value for °C, 52.

$$K = 52 + 273.15$$

Finally, calculate the desired value for K.

$$K = 325.15 \text{ K}$$

Atomic Structure and the Periodic Table

Today's primary model of the atom was proposed by scientist Niels Bohr. **Bohr's atomic model** consists of a nucleus, or core, which is made up of positively charged **protons** and neutrally charged **neutrons**. Neutrons are theorized to be in the nucleus with the protons to provide "balance" and stability to the protons at the center of the atom. More than 99 percent of the mass of an atom is found in the nucleus. Orbitals surrounding the nucleus contain negatively charged particles called **electrons**. Since the entire structure of an atom is too small to be seen with the unaided eye, an electron microscope is required for detection. Even with such magnification, the actual particles of the atom are not visible.

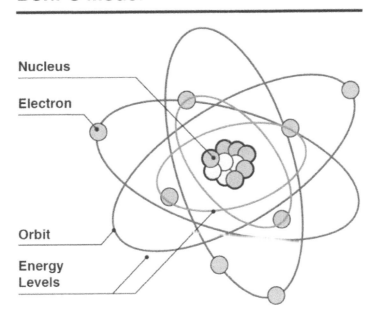

Anything that takes up space and has mass is considered **matter**. Matter is composed of atoms. An **atom** has an atomic number that is determined by the number of protons within the nucleus. **Properties**, which are observable characteristics of a substance, can be used to describe different substances. The **physical properties** of a substance can be observed without altering the identity or chemical composition of the substance. Density, solubility, malleability, odor, and luster are examples of physical properties. **Chemical properties**, on the other hand, can only be observed through a change in the chemical composition of the substance via a chemical reaction. Flammability, oxidative state, and reactivity with an acid are examples of chemical properties.

Some substances are made up of atoms, all with the same atomic number. Such a substance is called an **element**. Using their atomic numbers, elements are organized and grouped by similar properties in a chart called the **Periodic Table**. An example can be seen by going to this link or by scanning the QR code with your device:

apexprep.com/bonus/asvab

The sum of the total number of protons in an atom and the total number of neutrons in the atom provides the atom's **mass number**. Most atoms have a nucleus that is electronically neutral, and all atoms of one type have the same atomic number. There are some atoms of the same type that have a different mass number. The variation in the mass number is due to an imbalance of neutrons within the nucleus of the atoms. If atoms have this variance in neutrons, they are called **isotopes**. It is the different number of neutrons that gives such atoms a different mass number. For isotopes, the atomic number (determined by the number of protons) is the same, but the mass number (determined by adding the protons and neutrons) is different.

This is a result of there being a different number of neutrons.

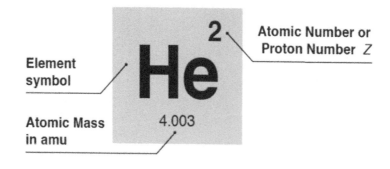

A concise method of arranging elements by atomic number, similar characteristics, and electron configurations in a tabular format was necessary to represent elements. This was originally organized by scientist Dmitri Mendeleev using the Periodic Table. The vertical lines on the Periodic Table are called **groups** and are sorted by similar chemical properties/characteristics, such as appearance and reactivity. This is observed in the shiny texture of metals, the softness of post-transition metals, and the high melting points of alkali earth metals. The horizontal lines on the Periodic Table are called **periods** and are arranged by electron valance configurations.

Elements are set by ascending atomic number, from left to right. The number of protons contained within the nucleus of the atom is represented by the atomic number. For example, hydrogen has one proton in its nucleus, so it has an atomic number of 1.

Since isotopes can have different masses within the same type of element, the atomic mass of an element is the average mass of all the naturally occurring atoms of that given element. **Atomic mass** is calculated by finding the relative abundance of isotopes that might be used in chemistry, or by adding the number of protons and neutrons of an atom together. For example, the mass number of one typical chlorine atom is 35: the nucleus has 17 protons (given by chlorine's atomic number of 17) and 18 neutrons. However, a large number of chlorine isotopes with a mass number of 37 exist in nature. These isotopes have 20 neutrons instead of 18 neutrons. The average of all the mass numbers turns out to be 35.5 amu, which is chlorine's atomic mass on the periodic table. In contrast, a typical carbon atom has a mass number of 12, and its atomic mass is 12.01 amu because there are not as many naturally occurring isotopes that raise the average number, as observed with chlorine.

Chemical Reactions

A **chemical reaction** is a process that involves a change in the molecular arrangement of a substance. Generally, one set of chemical substances, called the **reactants**, is rearranged into a different set of chemical substances, called the **products**, by the breaking and re-forming of bonds between atoms. In a chemical reaction, it is important to realize that no new atoms or molecules are introduced. The products are formed solely from the atoms and molecules that are present in the reactants. These can involve a change in state of matter as well. Making glass, burning fuel, and brewing beer are all examples of chemical reactions.

Generally, chemical reactions are thought to involve changes in positions of electrons with the breaking and re-forming of chemical bonds, without changes to the nucleus of the atoms. The four main types of chemical reactions are combination, decomposition, combustion, and oxidation/reduction reactions.

Combination
In **combination reactions**, two or more reactants are combined to form one more complex, larger product. The bonds of the reactants are broken, the elements are arranged, and then new bonds are formed between all the elements to form the product. It can be written as:

$$A + B \rightarrow C$$

A and B are the reactants and C is the product. An example of a combination reaction is the creation of iron (II) sulfide from iron and sulfur, which is written as:

$$8Fe + S_8 \rightarrow 8FeS$$

Decomposition

Decomposition reactions are almost the opposite of combination reactions. They occur when one substance is broken down into two or more products. The bonds of the first substance are broken, the elements are rearranged, and then the elements are bonded together in new configurations to make two or more molecules. These reactions can be written as:

$$C \rightarrow B + A$$

where C is the reactant and A and B are the products.

An example of a decomposition reaction is the electrolysis of water to make oxygen and hydrogen gas, which is written as:

$$2H_2O \rightarrow 2H_2 + O_2$$

Combustion

Combustion reactions are a specific type of chemical reaction that involves oxygen gas as a reactant. This mostly involves the burning of a substance. The combustion of hexane in air is one example of a combustion reaction. The hexane gas combines with oxygen in the air to form carbon dioxide and water. The reaction can be written as:

$$2C_6H_{14} + 17O_2 \rightarrow 12CO_2 + 14H_2O$$

Oxidation and Reduction

Oxidation/reduction (redox or half) reactions involve the oxidation of one species and the reduction of the other species in the reactants of a chemical equation. This can be seen through three main types of transfers.

The first type is through the **transfer of oxygen**. The reactant gaining an oxygen is the oxidizing agent, and the reactant losing an oxygen is the reduction agent.

For example, the oxidation of magnesium is as follows:

$$2\,Mg(s) + O_2(g) \rightarrow 2\,MgO(s)$$

The second type is through the **transfer of hydrogen**. The reactant losing the hydrogen is the oxidizing agent, and the other reactant is the reduction agent.

For example, the redox of ammonia and bromine results in nitrogen and hydrogen bromide due to bromine gaining a hydrogen as follows:

$$2\,NH_3 + 3\,Br_2 \rightarrow N_2 + 6\,HBr$$

The third type is through the loss of electrons from one species, known as the **oxidation agent**, and the gain of electrons to the other species, known as the **reduction agent**. For a reactant to become "oxidized," it must give up an electron.

For example, the redox of copper and silver is as follows:

$$Cu(s) + 2\,Ag^+(aq) \rightarrow Cu^{2+}(aq) + 2\,Ag(s)$$

14

It is also important to note that the oxidation numbers can change in a redox reaction due to the transfer of oxygen atoms. Standard rules for finding the oxidation numbers for a compound are listed below:

1. The oxidation number of a free element is always 0.

2. The oxidation number of a monatomic ion equals the charge of the ion.

3. The oxidation number of H is +1, but it is –1 when combined with less electronegative elements.

4. The oxidation number of O in compounds is usually –2, but it is –1 in peroxides.

5. The oxidation number of a Group 1 element in a compound is +1.

6. The oxidation number of a Group 2 element in a compound is +2.

7. The oxidation number of a Group 17 element in a binary compound is –1.

8. The sum of the oxidation numbers of all the atoms in a neutral compound is 0.

9. The sum of the oxidation numbers in a polyatomic ion is equal to the charge of the ion.

These rules can be applied to determine the oxidation number of an unknown component of a compound.

For example, what is the oxidation number of Cr in $CrCl_3$?

From rule 7, the oxidation number of Cl is given as –1. Since there are 3 chlorines in this compound, that would equal 3×-1 for a result of –3. According to rule 8, the total oxidation number of Cr must balance the total oxidation number of Cl, so Cr must have a total oxidation number equaling +3 ($-3 + +3 = 0$). There is only 1 Cr, so the oxidation number would be multiplied by 1, or the same as the total of +3, written as follows:

$$\overset{+3 \quad -1}{\underset{+3 \quad -3}{CrCl_3}}$$

Chemical Equations

Chemical equations describe how the molecules are changed when the chemical reaction occurs. For example, the chemical equation of the hexane combustion reaction is $2C_6H_{14} + 17O_2 \rightarrow 12CO_2 + 14H_2O$. The "+" sign on the left side of the equation indicates that those molecules are reacting with each other, and the arrow, "$\rightarrow$," in the middle of the equation indicates that the reactants are producing something else. The coefficient before a molecule indicates the quantity of that specific molecule that is present for the reaction. The subscript next to an element indicates the quantity of that element in each molecule. For the chemical equation to be balanced, the quantity of each element on both sides of the equation should be equal. For example, in the hexane equation above, there are twelve carbon elements, twenty-eight hydrogen elements, and thirty-four oxygen elements on each side of the equation. Even though they are part of different molecules on each side, the overall quantity is the same. The state of matter of the reactants and products can also be included in a chemical equation and would be written in parentheses next to each element as follows: gas (g), liquid (l), solid (s), and dissolved in water, or aqueous (aq).

Reaction Rates, Equilibrium, and Reversibility

The rate of a chemical reaction can be increased by adding a catalyst to the reaction. **Catalysts** are substances that lower the activation energy required to go from the reactants to the products of the reaction but are not consumed in the process. The **activation energy** of a reaction is the minimum amount of energy that is required to make the reaction move forward and change the reactants into the products. When catalysts are present, less energy is required to complete the reaction. For example, hydrogen peroxide will eventually decompose into two water molecules and one oxygen molecule. If potassium permanganate is added to the reaction, the decomposition happens at a much faster rate. Similarly, increasing the temperature or pressure in the environment of the reaction can increase the rate of the reaction. Higher temperatures increase the number of high-energy collisions that lead to the products. The same happens when increasing pressure for gaseous reactants, but not with solid or liquid reactants. Increasing the concentration of the reactants or the available surface area over which they can react also increases the rate of the reaction.

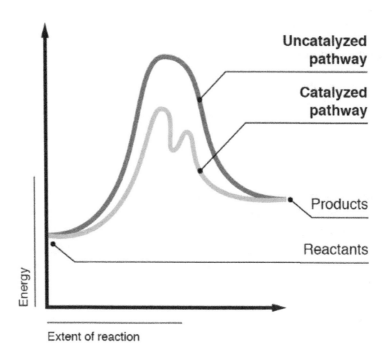

Many reactions are **reversible**, which means that the products can revert to the reactants under certain conditions. In such reactions, the arrow will be double-headed, indicating that the reaction can proceed in either direction. Reactions reach a state of **equilibrium** when the net concentration of the reactants and products are not changing. This does not mean that the reactions have stopped occurring, but that there is no overall change in either direction (forming more product from the reactants or the product undergoing the reverse reaction to re-form the reactants). Some number of reactions may be going on in both directions, but they cancel each other out so that there is no net change in concentrations on either side of the reaction arrow.

Solutions and Solubility

A **solution** is a homogenous mixture of two or more substances. Unlike heterogenous mixtures, in solutions, the **solute**, which is a substance that can be dissolved, is uniformly distributed throughout the **solvent**, which is the substance in which the solvent dissolves. For example, when 10 grams of table salt

16

(NaCl) is added to a 100 mL of room temperature water and then stirred until all the salt (the solute) has dissolved in the water (the solvent), a solution is formed. The dissolved salt, in the form of Na^+ and Cl^- ions, will be evenly distributed throughout the water. In this case, the solution is **diluted** because only a small amount of solute was dissolved in a comparatively large volume of solvent.

The saltwater solution would be said to be **concentrated** if a large amount of salt was added, stirred, and dissolved into the water, 30 grams, for example. When more solute is added to the solvent but even after stirring, some settles on the bottom without dissolving, the solution is **saturated**. For example, in 100 mL of room temperature water, about 35 grams of table salt can dissolve before the solution is saturated. Beyond this point—called the saturation point—any additional salt added will not readily dissolve. Sometimes, it is possible to temporarily dissolve excessive solute in the solvent, which creates a **supersaturated** solution. However, as soon as this solution is disturbed, the process of **crystallization** will begin and a solid will begin precipitating out of the solution.

It is often necessary, for example when working with chemicals or mixing acids and bases, to quantitatively determine the concentration of a solution, which is a more precise measure than using qualitative terms like diluted, concentrated, saturated, and supersaturated. The **molarity**, c, of a solution is a measure of its concentration; specifically, it is the number of moles of solute (represented by n in the formula) per liter of solution (V). Therefore, the following is the formula for calculating the molarity of a solution:

$$c = \frac{n}{V}$$

It is important to remember that the volume in the denominator of the equation above is in liters of solution, not solvent. Adding solute increases the volume of the entire solution, so the molarity formula accounts for this volumetric increase.

Factors Affecting Solubility

Solubility refers to a solvent's ability to dissolve a solute. Certain factors can affect solubility, such as temperature and pressure. Depending on the state of matter of the molecules in the solution, these factors may increase or decrease solubility. When most people think of solutions, they imagine a solid or liquid solute dissolved in a liquid solvent, like water. However, solutions can be composed on molecules in the other states of matter as well. The following are examples of solutions involving combinations of solids, liquids, and gasses:

- Dry air is solution of gasses, mainly nitrogen and oxygen, with carbon dioxide and others as well
- Vinegar is a solution of liquid acetic acid and liquid water
- Brass is a solution of solid copper and solid zinc
- Amalgams can be a solution of liquid mercury with a solid metal like gold
- Hummingbirds like sweet water, a solution of liquid water and solid sucrose (table sugar)

Increases in temperature, tend to increase a solute's solubility, except in the case of gasses wherein higher temperature reduce the solubility of gaseous solutions and gasses dissolved in water. However, the solubility of gasses in organic solvents increases with increases in temperature. Pressure is directly related to the solubility of gasses, but not of liquids or solids. Agitating, or stirring, a solution does not affect solubility, but it does increase the rate at which the solute dissolves.

The electronegativities, or **polarities**, of the solvent and solute determine whether the solute will readily dissolve in the solvent. The key to the potential for a solution to form and the solute to dissolve in the

17

solvent is "like dissolves like." If the solute is polar, it will dissolve in a polar solvent. If the solute is non-polar, it will dissolve in a non-polar solvent. Water is an example of a polar solvent, while benzene is an example of a non-polar solvent.

Stoichiometry

Stoichiometry uses proportions based on the principles of the conservation of mass and the conservation of energy. It deals with first balancing the chemical (and sometimes physical) changes in a reaction and then finding the ratios of the reactants used and what products resulted. Just as there are different types of reactions, there are different types of stoichiometry problems. Different reactions can involve mass, volume, or moles in varying combinations. The steps to solve a stoichiometry problem are to first, balance the equation; next, find the number of total products; and finally, calculate the desired information regarding molar mass, percent yield, etc.

The **molar mass** of any substance is the measure of the mass of one mole of that substance. For pure elements, the molar mass is also referred to as the **atomic mass unit (amu)** for that substance. In compounds, this is calculated by adding the molar mass of each substance in the compound. For example, the molar mass of carbon can be found on the Periodic Table as 12.01 g/mol, while finding the molar mass of water (H_2O) requires a bit of calculation.

$$
\begin{array}{lll}
\text{the molar mass of hydrogen} = & 1.01 \times 2 & = \quad 2.02 \\
+ \text{ the molar mass of oxygen} = & 16.0 & = +\underline{16.00} \\
& & \quad 18.02 \text{ g/mol}
\end{array}
$$

To determine the **percent composition** of a compound, the individual molar masses of each component need to be divided by the total molar mass of the compound and then multiplied by 100. For example, to find the percent composition of carbon dioxide (CO_2) first requires the calculation of the molar mass of CO_2.

$$
\begin{array}{lll}
\text{the molar mass of carbon} = & 12.01 & = \quad 12.01 \\
+ \text{ the molar mass of oxygen} = & 16.0 \times 2 & = +\underline{32.00} \\
& & \quad 44.01 \text{ g/mol}
\end{array}
$$

Next, take each individual mass, divide by the total mass of the compound, and then multiply by 100 to get the percent composition of each component.

$$\frac{12.01 \text{ g/mol}}{44.01 \text{ g/mol}} = 0.2729 \times 100 = 27.29\% \text{ carbon}$$

$$\frac{32.00 \text{ g/mol}}{44.01 \text{ g/mol}} = 0.7271 \times 100 = 72.71\% \text{ oxygen}$$

A quick check in the addition of the percentages should always total 100%.

If an example provides the basis for an equation, the equation will first need to be balanced to calculate any proportions. For example, if 15 g of C_2H_6 reacts with 64 g of O_2, how many grams of CO_2 will be formed?

First, write the chemical equation:

$$C_2H_6 + O_2 \rightarrow CO_2 + H_2O$$

Next, balance the equation:

$$2\,C_2H_6 + 7\,O_2 \rightarrow 4\,CO_2 + 6\,H_2O$$

Then, calculate the desired amount based on the beginning information of 15 g of C_2H_6:

$$15\,\text{g}\,C_2H_6 \times \frac{1\text{ mole }C_2H_6}{30\text{ g }C_2H_6} \times \frac{4\text{ moles }CO_2}{2\text{ moles }C_2H_6} \times \frac{44\text{ g }CO_2}{1\text{ mole }CO_2} = 44\,\text{g}\,CO_2$$

To check that this would be the smaller amount (or how much until one of the reactants is used up, thus ending the reaction), the calculation would need to be done for the 64 g of O_2:

$$64\,\text{g}\,O_2 \times \frac{1\text{ mole }O_2}{32\text{ g }O_2} \times \frac{4\text{ moles }CO_2}{7\text{ moles }O_2} \times \frac{44\text{ g }CO_2}{1\text{ mole }CO_2} = 50.5\,\text{g}\,CO_2$$

The yield from the C_2H_6 is smaller, so it would be used up first, ending the reaction. This calculation would determine the maximum amount of CO_2 that could possibly be produced.

Acids and Bases

If something has a sour taste, it is considered acidic, and if something has a bitter taste, it is considered basic. Acids and bases are generally identified by the reaction they have when combined with water. An **acid** will increase the concentration of hydrogen ions (H^+) in water, and a **base** will increase the concentration of hydroxide ions (OH^-). Other methods of identification with various indicators have been designed over the years.

To better categorize the varying strength levels of acids and bases, the pH scale is employed. The **pH scale** is a logarithmic (base 10) grading applied to acids and bases according to their strength. The pH scale contains values from 0 through 14 and uses 7 as neutral. If a solution registers below a 7 on the pH scale, it is considered an acid. If a solution registers higher than a 7, it is considered a base. To perform a quick test on a solution, litmus paper can be used. A base will turn red litmus paper blue, and an acid will turn blue litmus paper red. To gauge the strength of an acid or base, a test using phenolphthalein can be administered. An acid will turn red phenolphthalein to colorless, and a base will turn colorless phenolphthalein to pink. As demonstrated with these types of tests, acids and bases neutralize each other. When acids and bases react with one another, they produce salts (also called **ionic substances**).

Acids and bases have varying strengths. For example, if an acid completely dissolves in water and ionizes, forming an H^+ and an anion, it is considered a strong acid. There are only a few common strong acids, including sulfuric (H_2SO_4), hydrochloric (HCl), nitric (HNO_3), hydrobromic (HBr), hydroiodic (HI), and perchloric ($HClO_4$). Other types of acids are considered weak.

An easy way to tell if something is an acid is by looking for the leading "H" in the chemical formula.

A base is considered strong if it completely dissociates into the cation of OH^-, including sodium hydroxide (NaOH), potassium hydroxide (KOH), lithium hydroxide (LiOH), cesium hydroxide (CsOH), rubidium hydroxide (RbOH), barium hydroxide ($Ba(OH)_2$), calcium hydroxide ($Ca(OH)_2$), and strontium hydroxide

(Sr(OH)$_2$). Just as with acids, other types of bases are considered weak. An easy way to tell if something is a base is by looking for the "OH" ending on the chemical formula.

In pure water, autoionization occurs when a water molecule (H$_2$O) loses the nucleus of one of the two hydrogen atoms to become a hydroxide ion (OH$^-$). The nucleus then pairs with another water molecule to form hydronium (H$_3$O$^+$). This autoionization process shows that water is **amphoteric**, which means it can react as an acid or as a base.

Pure water is considered neutral, but the presence of any impurities can throw off this neutral balance, causing the water to be slightly acidic or basic. This can include the exposure of water to air, which can introduce carbon dioxide molecules to form carbonic acid (H$_2$CO$_3$), thus making the water slightly acidic. Any variation from the middle of the pH scale (7) indicates a non-neutral substance.

Nuclear Chemistry

Nuclear chemistry (also referred to as **nuclear physics**) deals with interactions within the nuclei of atoms. This differs from typical chemical reactions, which involve interactions with the electrons of atoms. If the nucleus of an atom is unstable, it emits radiation as it releases energy resulting from changes in the nucleus. This instability often occurs in isotopes of an element. An **isotope** is formed when the nucleus of an atom has the same number of protons but a different number of neutrons. This difference in mass causes a heavy, unstable condition in the nucleus.

According to quantum theory, there is no way to precisely predict when an atom will decay, but the decay of a collection of atoms in a substance can be predicted by their collective half-life. **Half-life** is used to calculate the time it takes for nuclei in atoms of a radioactive substance to have undergone radioactive decay (in which they emit particles and energy).

There are three primary types of nuclear decay occurring in an atom with an unstable nucleus: alpha, beta, and gamma.

In **alpha decay**, an atom will emit two protons and two neutrons from its nucleus. This emission is in the form of a "bundle" and is called an **alpha particle**. Occurring mainly in larger, heavier atoms, alpha decay causes the atom's proton count to drop by two, thus resulting in the creation of a new element. Alpha radiation is extremely weak and can be blocked by something as thin as a piece of paper.

When the neutron of an atom emits an electron and causes the electron count to increase, the proton count of the nucleus is also increased. This action creates a new element and emits beta radiation in the process. **Beta radiation** is slightly more dangerous than alpha radiation, but it can be blocked by heavy materials such as aluminum or even wood.

The most dangerous type of radiation results from gamma decay. **Gamma decay** does not alter the mass or the charge of an atom. Gamma radiation is emitted along with an alpha or beta particle. It is extremely dangerous and can only be blocked by lead. For example, iodine is a stable element, often used in nuclear medicine. Iodine's atomic number is 53, and its atomic mass is 127 amu. When an isotope of iodine, with atomic number 53 and atomic mass of 131 amu (due to an excess of four neutrons), is used in the human body, it can be seen in the thyroid as a radioactive tracer.

When naturally occurring radioactive elements decay, they follow a radioactive decay series, which starts as one element that decays into a second element, which then decays into a third element, and so on until a stable element is finally reached. There are three naturally occurring radioactive decay series. Each of these series begins with either uranium-235, uranium-238, or thorium. For example, uranium-238 decays into astatine, bismuth, lead, polonium, protactinium, radium, radon, thallium, and thorium before finally becoming a more stable element. An artificial radioactive series begins with the artificially made element of neptunium and decays into elements such as polonium and americium on its way to becoming more stable.

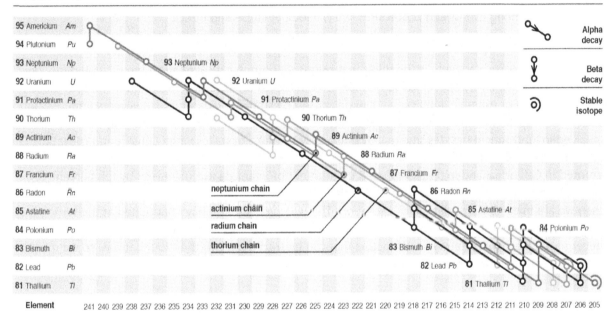

When naturally occurring radioactive elements decay, they follow a radioactive decay series, which starts as one element, decays into a second element, decays into a third element, and so on until a stable element is finally reached. There are three naturally occurring radioactive decay series. Each of these series begins with either uranium-235, uranium-238, or thorium. For example, uranium-238 decays into astatine, bismuth, lead, polonium, protactinium, radium, radon, thallium, and thorium before finally becoming a more stable element. An artificial radioactive series begins with the artificially made element of neptunium and decays into elements such as polonium and americium on its way to becoming more stable.

Isotopes can be created through separation and synthesis. There are approximately two hundred radioisotopes; most of them are artificially created. Bombarding the nucleus with nuclear particles or nuclei changes the nucleus of an element. This method is called **transmutation** and has resulted in the formation of new, artificial elements.

The application of half-life knowledge can be seen through the carbon cycle used in carbon dating. Some elements have been estimated to be as old as our universe, and studying the position of something such as carbon in its half-life decay cycle can provide an accurate age of a substance containing carbon. Carbon-14 is utilized for this specific purpose.

The understanding of nuclear chemistry is important for scientists to design and utilize radioactive drugs for treatments and diagnostic techniques. If an isotope is radioactive, it can easily be detected as a contrast agent in the human body, enabling medical professionals to view and diagnose issues that could not be observed without contrast or through physical exams or x-rays. When radioactive tracers are introduced, they can aid in the detection of how systems work. Since the amount of radiation used is small, it can help create a map of its path through a system without causing any disruption to that system.

The radioactive materials can also provide the necessary backdrop or contrast for reading emissions from samples by spectroscopes.

Exposure to radiation has varying effects on the human body. Medical treatments can utilize short exposures to radiation, and damage can be localized to specific, targeted areas, as in treatments for cancer. Extended exposure can cause chromosome damage through breaking the chemical bonds of DNA. Prolonged exposure to radiation can even cause cancer by the mutation and killing of cells and through diminishing the body's ability to produce cells and heal. Overexposure to radiation can result in death.

Relationships Among Events, Objects, and Procedures

When we determine relationships among events, objects, and procedures, we are better able to understand the world around us and make predictions based on that understanding. With regards to relationships among events and procedures, we will look at cause and effect.

Cause

The **cause** of a particular event is the thing that brings it about. A causal relationship may be partly or wholly responsible for its effect, but sometimes it's difficult to tell whether one event is the sole cause of another event. For example, lung cancer can be caused by smoking cigarettes. However, sometimes lung cancer develops even though someone does not smoke, and that tells us that there may be other factors involved in lung cancer besides smoking. It's also easy to make mistakes when considering causation. One common mistake is mistaking correlation for causation. For example, say that in the year 2008 a volcano erupted in California, and in that same year, the number of infant deaths increased by ten percent. If we automatically assume, without looking at the data, that the erupting volcano *caused* the infant deaths, then we are mistaking correlation for causation. The two events might have happened at the same time, but that does not necessarily mean that one event caused the other. Relationships between events are never absolute; there are a myriad of factors that can be traced back to their foundations, so we must be thorough with our evidence in proving causation.

Effect

An **effect** is the result of something that occurs. For example, the Nelsons have a family dog. Every time the dog hears thunder, the dog whines. In this scenario, the thunder is the cause, and the dog's whining is the effect. Sometimes a cause will produce multiple effects. Let's say we are doing an experiment to see what the effects of exercise are in a group of people who are not usually active. After about four weeks, the group experienced weight loss, rise in confidence, and higher energy. We start out with a cause: exercising more. From that cause, we have three known effects that occurred within the group: weight loss, rise in confidence, and higher energy. Cause and effect are important terms to understand when conducting scientific experiments.

Physics

Nature of Motion

Cultures have been studying the movement of objects since ancient times. These studies have been prompted by curiosity and sometimes by necessity. On Earth, items move according to guidelines and have motion that is predictable. To understand why an object moves along its path, it is important to understand what role forces have on influencing its movements. The term **force** describes an outside influence on an object. Force does not have to refer to something imparted by another object. Forces can act upon objects by touching them with a push or a pull, by friction, or without touch like a magnetic force or even gravity. Forces can affect the motion of an object.

To study an object's motion, it must be located and described. When locating an object's position, it can help to pinpoint its location relative to another known object. Comparing an object with respect to a known object is referred to as **establishing a frame of reference**. If the placement of one object is known, it is easier to locate another object with respect to the position of the original object.

Motion can be described by following specific guidelines called **kinematics**. Kinematics use mechanics to describe motion without regard to the forces that are causing such motions. Specific equations can be used when describing motions; these equations use time as a frame of reference. The equations are based on the change of an object's position (represented by x), over a change in time (represented by Δt). This describes an object's velocity, which is measured in meters/second (m/s) and described by the following equation:

$$v = \frac{\Delta x}{\Delta t} = \frac{x_f - x_i}{\Delta t}$$

Velocity is a **vector quantity**, meaning it measures the magnitude (how much) and the direction that the object is moving. Both components are essential to understanding and predicting the motion of objects. The scientist Isaac Newton did extensive studies on the motion of objects on Earth and came up with three primary laws to describe motion:

Law 1: An object in motion tends to stay in motion unless acted upon by an outside force. An object at rest tends to stay at rest unless acted upon by an outside force (also known as the **law of inertia**).

For example, if a book is placed on a table, it will stay there until it is moved by an outside force.

Law 2: The force acting upon an object is equal to the object's mass multiplied by its acceleration (also known as $F = ma$).

For example, the amount of force acting on a bug being swatted by a flyswatter can be calculated if the mass of the flyswatter and its acceleration are known. If the mass of the flyswatter is 0.3 kg and the acceleration of its swing is 2.0 m/s², the force of its swing can be calculated as follows:

$$m = 0.3 \text{ kg}$$
$$a = 2.0 \text{ m/s}^2$$
$$F = m \times a$$
$$F = (0.3) \times (2.0)$$
$$F = 0.6 \text{ N}$$

Law 3: For every action, there is an equal and opposite reaction.

For example, when a person claps their hands together, the right hand feels the same force as the left hand, as the force is equal and opposite.

Another example is if a car and a truck run head-on into each other, the force experienced by the truck is equal and opposite to the force experienced by the car, regardless of their respective masses or velocities. The ability to withstand this amount of force is what varies between the vehicles and creates a difference in the amount of damage sustained.

Newton used these laws to describe motion and derive additional equations for motion that could predict the position, velocity, acceleration, or time for objects in motion in one and two dimensions. Since all of Newton's work was done on Earth, he primarily used Earth's gravity and the behavior of falling objects to design experiments and studies in free fall (an object subject to Earth's gravity while in flight). On Earth, the acceleration due to the force of gravity is measured at 9.8 meters per second^2 (m/s^2). This value is the same for anything on the Earth or within Earth's atmosphere.

Acceleration

Acceleration is the change in velocity over the change in time. It is given by the following equation:

$$a = \frac{\Delta v}{\Delta t} = \frac{v_f - v_i}{\Delta t}$$

Since velocity is the change in position (displacement) over a change in time, it is necessary for calculating an acceleration. Both are vector quantities, meaning they have magnitude and direction (or some amount in some direction). Acceleration is measured in units of distance over time^2 (meters/second^2 or m/s^2 in metric units).

For example, what is the acceleration of a vehicle that has an initial velocity of 35 m/s and a final velocity of 10 m/s over 5.0 s?

Using the givens and the equation:

$$a = \frac{\Delta v}{\Delta t} = \frac{v_f - v_i}{\Delta t}$$

$$V_f = 10 \text{ m/s}$$

$$V_i = 35 \text{ m/s}$$

$$\Delta t = 5.0 \text{ s}$$

$$a = \frac{10 - 35}{5.0} = \frac{-25}{5.0} = -5.0 \text{ m/s}^2$$

The vehicle is decelerating at –5.0 m/s^2.

If an object is moving with a constant velocity, its velocity does not change over time. Therefore, it has no (or 0) acceleration.

It is common to misuse vector terms in spoken language. For example, people frequently use the term "speed" in situations where the correct term would be "velocity." However, the difference between velocity and speed is not just that velocity must have a direction component with it. Average velocity and average speed are looking at two different distances as well. Average speed is calculated simply by dividing the total distance covered by the time it took to travel that distance. If someone runs four miles along a straight road north and then makes a 90-degree turn to the right and runs another three miles down that straight road east (such that the runner's route of seven miles makes up two sides of a rectangle) in seventy minutes, the runner's average speed was 6 miles per hour (one mile covered every ten minutes). Using the same course, the runner's average velocity would be about 4.29 miles per hour northeast.

Why is the magnitude less in the case of velocity? Velocity measures the change in position, or *displacement,* which is the shortest line between the starting point and ending point. Even if the path between these two points is serpentine or meanders all over the place racking up a great distance, the displacement is still just the shortest straight path between the change in the position of the object. In the case of the runner, the "distance" used to calculate velocity (the displacement) is the hypotenuse of the triangle that would connect the two side lengths at right angles to one another.

Using basic trigonometric ratios, we know this distance is 5 miles (since the lengths of the other two legs are 3 miles and 4 miles and the Pythagorean Theorem says that $a^2 + b^2 = c^2$). Therefore, although the distance the runner covered was seven miles, their displacement was only five miles. Average velocity is thus calculated by taking the total time (70 minutes) and dividing it by the displacement (5 miles northeast). Therefore, to calculate average velocity of the runner, 70 minutes is divided by 5 miles, so each mile of displacement to the northeast was covered in 14 minutes. To find this rate in miles per hour, 60 minutes is divided by 14, to get 4.29 miles per hour northeast.

Another misconception is if something has a negative acceleration, it must be slowing down. If the change in position of the moving object is in a negative direction, it could have a negative velocity. If the acceleration is in the same direction as this negative velocity, it would be increasing the velocity in the negative direction, thus resulting in the object increasing in velocity.

For example, if west is designated to be a negative direction, a car increasing in speed to the west would have a negative velocity. Since it is increasing in speed, it would be accelerating in the negative direction, resulting in a negative acceleration.

Another common misconception is if a person is running around an oval track at a constant velocity, they would have no (or 0) acceleration because there is no change in the runner's velocity. This idea is incorrect because the person is changing direction the entire time they are running around the track, so there would be a change in their velocity, therefore; the runner would have an acceleration.

One final point regarding acceleration is that it can result from the force a rotating body exerts toward its center. For planets and other massive bodies, it is called **gravity**. This type of acceleration can also be utilized to separate substances, as in a centrifuge.

Projectile Motion

When objects are launched or thrown into the air, they exhibit what is called **projectile motion**. This motion takes a parabolic (or arced) path as the object rises and/or falls with the effect of gravity. In sports, if a ball is thrown across a field, it will follow a path of projectile motion. Whatever angle the object leaves the horizon is the same angle with which it will return to the horizon. The height the object achieves is referred to as the y-**component**, and the distance along the horizon the object travels is referred to as the x-**component**. To maximize the horizontal distance an object travels, it should be launched at a 45-degree angle relative to the horizon.

The following shows the range, or x-distance, an object can travel when launched at various angles:

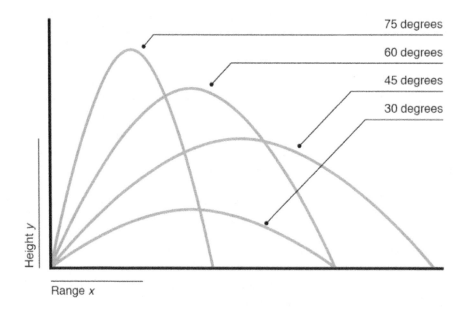

If something is traveling through the air without an internal source of power or any extra external forces acting upon it, it will follow these paths. All projectiles experience the effects of gravity while on Earth; therefore, they will experience a constant acceleration of 9.8 m/s^2 in a downward direction. While on its path, a projectile will have both a horizontal and vertical component to its motion and velocity. At the launch, the object has both vertical and horizontal velocity. As the object increases in height, the y-component of its velocity will diminish until the very peak of the object's path.

At the peak, the y-component of the velocity is zero. Since the object still has a horizontal, or x-component, to its velocity, it would continue its motion. There is also a constant acceleration of 9.8 m/s^2 acting in the downward direction on the object the entire time, so at the peak of the object's height, the velocity would then switch from zero to being pulled down with the acceleration. The entire path of the object takes a specific amount of time, based on the initial launch angle and velocity. The time and distance traveled can be calculated using the kinematic equations of motion. The time it takes the object to reach its maximum height is exactly half of the time for the object's entire flight.

Similar motion is exhibited if an object is thrown from atop a building, bridge, cliff, etc. Since it would be starting at its maximum height, the motion of the object would be comparable to the second half of the path in the above diagram.

Friction

As previously stated with Newton's laws, forces act upon objects on the Earth. If an object is resting on a surface, the effect of gravity acting upon its mass produces a force referred to as **weight**. This weight touches the surface it is resting on, and the surface produces a normal force perpendicular to this surface. If an outside force acts upon this object, its movement will be resisted by the surfaces rubbing on each other.

Friction is the term used to describe the force that opposes motion, or the force experienced when two surfaces interact with each other. Every surface has a specific amount with which it resists motion, called a

coefficient of friction. The coefficient of friction is a proportion calculated from the force of friction divided by the **normal force** (force produced perpendicular to a surface).

$$\mu_s = \frac{F_s}{F_N}$$

There are different types of friction between surfaces. If something is at rest, it has a **static** (non-moving) friction. It requires an outside force to begin its movement. The coefficient of static friction for that material multiplied by the normal force would need to be greater than the force of static friction to get the object moving. Therefore, the force required to move an object must be greater than the force of static friction:

$$F_s \leq \mu_s \times F_N$$

Once the object is in motion, the force required to maintain this movement only needs to be equivalent to the value of the force of kinetic (moving) friction. To calculate the force of **kinetic** friction, simply multiply the coefficient of kinetic friction for that surface by the normal force:

$$F_k = \mu_k \times F_N$$

The force required to start an object in motion is larger than the force required to continue its motion once it has begun:

$$F_s \geq F_k$$

Friction not only occurs between solid surfaces; it also occurs in air and liquids. In air, it is called **air resistance**, or drag, and in water, it is called **viscosity**.

For example, what would the coefficient of static friction be if a 5.0 N force was applied to push a 20 kg crate, from rest, across a flat floor?

First, the normal force could be found to counter the force from the weight of the object, which would be the mass multiplied by gravity:

$$F_N = mass \times gravity$$

$$F_N = 20 \text{ kg} \times 9.8 \, \frac{\text{m}}{\text{s}^2}$$

$$F_N = 196 \text{ N}$$

Next, the coefficient of static friction could be found by dividing the frictional force by the normal force:

$$\mu_s = \frac{F_s}{F_N}$$

$$\mu_s = \frac{5.0 \text{ N}}{196 \text{ N}}$$

$$\mu_s = 0.03$$

Since it is a coefficient, the units cancel out, so the solution is unitless. The coefficient of static friction should also be less than 1.0.

Rotation

An object moving on a circular path has **momentum** (a measurement of an object's mass and velocity in a direction); for circular motion, it is called **angular momentum**, and this value is determined by rotational inertia, rotational velocity, and the distance of the mass from the axis of rotation, or the center of rotation.

If objects are exhibiting circular motion, they are demonstrating the conservation of angular momentum. The angular momentum of a system is always constant, regardless of the placement of the mass. As stated above, the rotational inertia of an object can be affected by how far the mass of the object is placed with respect to the center of rotation (axis of rotation). A larger distance between the mass and the center of rotation means a slower rotational velocity. Reversely, if a mass is closer to the center of rotation, the rotational velocity increases. A change in the placement of the mass affects the value of the rotational velocity, thus conserving the angular momentum. This is true as long as no external forces act upon the system.

For example, if an ice skater is spinning on one ice skate and extends their arms out (or increases the distance between the mass and the center of rotation), a slower rotational velocity is created. When the skater brings the arms in close to the body (or lessens the distance between the mass and the center of rotation), their rotational velocity increases, and he or she spins much faster. Some skaters extend their arms straight up above their head, which causes the axis of rotation to extend, thus removing any distance between the mass and the center of rotation and maximizing the rotational velocity.

Consider another example. If a person is selecting a horse on a merry-go-round, the placement of their selection can affect their ride experience. All the horses are traveling with the same rotational speed, but to travel along the same plane as the merry-go-round, a horse closer to the outside will have a greater linear speed due to being farther away from the axis of rotation. Another way to think of this is that an outside horse must cover more distance than a horse near the inside to keep up with the rotational speed of the merry-go-round platform. Based on this information, thrill seekers should always select an outer horse to experience a greater linear speed.

Uniform Circular Motion

When an object exhibits circular motion, its motion is centered around an axis. An **axis** is an invisible line on which an object can rotate. This type of motion is most easily observed on a toy top. There is a point (or rod) through the center of the top on which the top can be observed to be spinning. This is also called the axis. An axis is the location about which the mass of an object or system would rotate if free to spin.

In the instance of utilizing a lever to lift an object, it can be helpful to calculate the amount of force needed at a specific distance, applied perpendicular to the axis of motion, to calculate the torque, or circular force, necessary to move something. This is also employed when using a wrench to loosen a bolt. The equation for calculating the force in a circular direction, or perpendicular to an axis, is as follows:

$$Torque = F_\perp \times distance\ of\ lever\ arm\ from\ the\ axis\ of\ rotation$$

$$\tau = F_\perp \times d$$

For example, what torque would result from a 20 N force being applied to a lever 5 meters from its axis of rotation?

$$\tau = 20\,\text{N} \times 5\,\text{m}$$

$$\tau = 100\,\text{N} \times \text{m}$$

The amount of torque would be 100 N×m. The units would be Newton meters because it is a force applied at a distance away from the axis of rotation.

When objects move in a circle by spinning on their own axis, or because they are tethered around a central point (also considered an axis), they exhibit circular motion. **Circular motion** is similar in many ways to linear (straight line) motion; however, there are some additional facts to note. When an object spins or rotates on or around an axis, a force that feels like it is pushing out from the center of the circle is created. The force is pulling into the center of the circle. A reactionary force is what is creating the feeling of pushing out. The inward force is the real force, and this is called **centripetal force**. The outward, or reactionary, force is called **centrifugal force**. The reactionary force is not the real force; it just feels like it is there. This can also be referred to as a **fictional force**. The true force is the one pulling inward, or the centripetal force. The terms centripetal and centrifugal are often mistakenly interchanged.

For example, the method a traditional-style washing machine uses to spin a load of clothes to expunge the water from the load is to spin the machine barrel in a circle at a high rate of speed. During this spinning, the centripetal force is pulling in toward the center of the circle. At the same time, the reactionary force to the centripetal force is pressing the clothes up against the outer sides of the barrel, which expels the water out of the small holes that line the outer wall of the barrel.

Kinetic Energy and Potential Energy

There are two main types of energy. The first type is called **potential energy** (or gravitational potential energy), and it is stored energy, or energy due to an object's height from the ground.

The second type is called **kinetic energy**. Kinetic energy is the energy of motion. If an object is moving, it will have some amount of kinetic energy.

For example, if a roller-coaster car is sitting on the track at the top of a hill, it would have all potential energy and no kinetic energy. As the roller coaster travels down the hill, the energy converts from potential energy into kinetic energy. At the bottom of the hill, where the car is traveling the fastest, it would have all kinetic energy and no potential energy.

Another measure of energy is the **total mechanical energy** in a system. This is the sum (or total) of the potential energy plus the kinetic energy of the system. The total mechanical energy in a system is always conserved. The amounts of the potential energy and kinetic energy in a system can vary, but the total mechanical energy in a situation would remain the same.

The equation for the mechanical energy in a system is as follows:

$$ME = PE + KE$$

$$(Mechanical\ Energy\ =\ Potential\ Energy\ +\ Kinetic\ Energy)$$

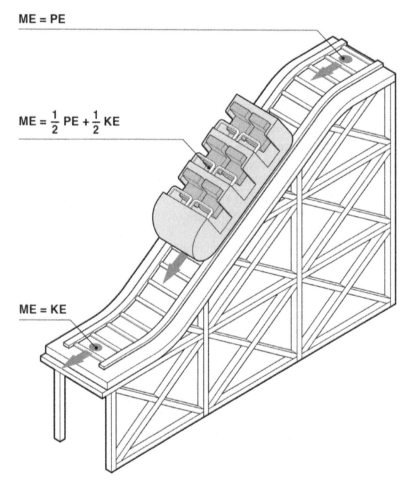

Energy can transfer or change forms, but it cannot be created or destroyed. This transfer can take place through waves (including light waves and sound waves), heat, impact, etc.

There is a fundamental law of thermodynamics (the study of heat and movement) called **conservation of energy**. This law states that energy cannot be created or destroyed, but rather energy is transferred to different forms involved in a process. For instance, a car pushed beginning at one end of a street will not continue down that street forever; it will gradually come to a stop some distance away from where it was originally pushed. This does not mean the energy has disappeared or has been exhausted; it means the energy has been transferred to different mediums surrounding the car. Some of the energy is dissipated by the frictional force from the road on the tires, the air resistance from the movement of the car, the sound from the tires on the road, and the force of gravity pulling on the car. Each value can be calculated in a number of ways, including measuring the sound waves from the tires, the temperature change in the tires, the distance moved by the car from start to finish, etc. It is important to understand that many processes factor into such a small situation, but all situations follow the conservation of energy.

Just like the earlier example, the roller coaster at the top of a hill has a measurable amount of potential energy; when it rolls down the hill, it converts most of that energy into kinetic energy. There are still additional factors such as friction and air resistance working on the coaster and dissipating some of the energy, but energy transfers in every situation.

There are six basic machines that utilize the transfer of energy to the advantage of the user. These machines function based on an amount of energy input from the user and accomplish a task by distributing the energy for a common purpose. These machines are called **simple machines** and include the lever, pulley, wedge, inclined plane, screw, and wheel and axle.

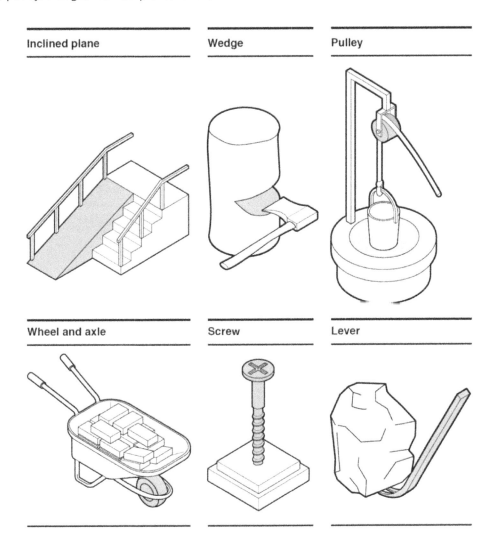

The use of simple machines can help by requiring less force to perform a task with the same result. This is referred to as a **mechanical advantage**.

For example, if a father is trying to lift his child into the air with his arms to pick an apple from a tree, it would require less force to place the child on one end of a teeter totter and push the other end of the teeter totter down to elevate the child to the same height to pick the apple. In this example, the teeter totter is a lever.

Linear Momentum and Impulse

The motion of an object can be expressed as momentum. This is a calculation of an object's mass times its velocity. **Momentum** can be described as the amount an object will continue moving along its current course. Momentum in a straight line is called **linear momentum**. Just as energy can be transferred and conserved, so can momentum.

Momentum is denoted by the letter p and calculated by multiplying an object's mass by its velocity.

$$p = m \times v$$

For example, if a car and a truck are moving at the same velocity (25 meters per second) down a highway, they will not have the same momentum because they do not have the same mass. The mass of the truck (3500 kg) is greater than that of the car (1,000kg); therefore, the truck will have more momentum. In a head-on collision, the truck's momentum is greater than the car's, and the truck will cause more damage to the car than the car will to the truck. The equations to compare the momentum of the car and the truck are as follows:

$p_{truck} = mass_{truck} \times velocity_{truck}$ $p_{car} = mass_{car} \times velocity_{car}$

$p_{truck} = 3{,}500 \text{ kg} \times 25 \text{ m/s}$ $p_{car} = 1{,}000 \text{ kg} \times 25 \text{ m/s}$

$p_{truck} = 87{,}500 \text{ N}$ $p_{car} = 25{,}000 \text{ N}$

The momentum of the truck is greater than that of the car.

The amount of force during a length of time creates an impulse. This means if a force acts on an object during a given amount of time, it will have a determined impulse. However, if the length of time can be extended, the force will be less due to the conservation of momentum.

For a car crash, the total momentum of each car before the collision would need to equal the total momentum of the cars after the collision. There are two main types of collisions: elastic and inelastic. For the example with a car crash, in an elastic collision, the cars would be separate before the collision, and they would remain separated after the collision. In the case of an inelastic collision, the cars would be separate before the collision, but they would be stuck together after the collision. The only difference would be in the way the momentum is calculated.

For elastic collisions:

$$total\ momentum_{before} = total\ momentum_{after}$$

$$(mass_{car\ 1} \times velocity_{car\ 1}) + (mass_{car\ 2} \times velocity_{car\ 2})$$
$$= (mass_{car\ 1} + mass_{car\ 2}) \times velocity_{car\ 1\ \&\ car\ 2}$$

The damage from an impact can be lessened by extending the time of the actual impact. This is called the measure of the impulse of a collision. It can be calculated by multiplying the change in momentum by the amount of time involved in the impact.

$$I = change\ in\ momentum \times time$$

$$I = \Delta p \times time$$

If the time is extended, the force (or change in momentum) is decreased. Conversely, if the time is shortened, the force (or change in momentum) is increased. For example, when catching a fast baseball, it helps soften the blow of the ball to follow through, or cradle the catch. This technique is simply extending the time of the application of the force of the ball, so the impact of the ball does not hurt the hand.

For example, if martial arts experts want to break a board by executing a chop from their hands, they need to exert a force on a small point on the board, extremely quickly. If they slow down the time of the impact from the force of their hands, they will probably injure their striking hand and not break the board.

Often, law enforcement officials will use rubber bullets instead of regular bullets to apprehend a criminal. The benefit of the rubber bullet is that the elastic material of the bullet bounces off the target but hits the target with nearly the same momentum as a regular bullet. Since the length of time the rubber bullet is in contact with the target is decreased, the amount of force from the bullet is increased. This method can knock a subject off their feet by the large force and the short time of the impact without causing any lasting harm to the individual. The difference in the types of collisions is noted through the rubber bullet bouncing off the individual, so both the bullet and the subject are separate before the collision and separate after the collision. With a regular bullet, the bullet and subject are separate before the collision, but a regular bullet would most likely not be separated by the subject after the collision.

Universal Gravitation

Every object in the universe that has mass causes an attractive force to every other object in the universe. The amount of attractive force depends on the masses of the two objects in question and the distance that separates the objects. This is called the **law of universal gravitation** and is represented by the following equation:

$$F = G \frac{m_1 m_2}{r^2}$$

In this equation, the force, F, between two objects, m_1 and m_2, is indirectly proportional to the square of the distance separating the two objects. A general gravitational constant G ($6.67 \times 10^{-11} \frac{N \times m^2}{kg^2}$) is multiplied by the equation. This constant is quite small, so for the force between two objects to be noticeable, they must have sizable masses.

To better understand this on a large scale, a prime representation could be viewed by satellites (planets) in the solar system and the effect they have on each other. All bodies in the universe have an attractive force between them. This is closely seen by the relationship between the Earth and the Moon. The Earth and the Moon both have a gravitational attraction that affects each other. The Moon is smaller in mass than the Earth; therefore, it will not have as big of an influence as the Earth has on it. The attractive force from the Moon is observed by the systematic push and pull on the water on the face of the Earth by the rotations the Moon makes around the Earth.

The tides in oceans and lakes are caused by the Moon's gravitational effect on the Earth. Since the Moon and the Earth have an attractive force between them, the Moon pulls on the side of the Earth closest to the Moon, causing the waters to swell (high tide) on that side and leave the ends 90 degrees away from the Moon, causing a low tide there. The water on the side of the Earth farthest from the Moon experiences the least amount of gravitational attraction so it collects on that side in a high tide.

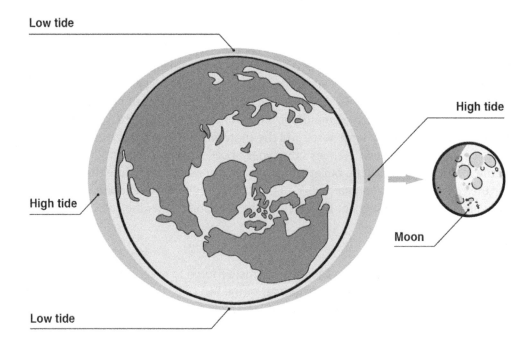

The universal law of gravitation is taken primarily from the works of Johannes Kepler and his laws of planetary motion. These include the fact that the paths of the orbits of the planets are not perfect circles, but ellipses, around the sun. The area swept out between the planet and the sun is equal at every point in the orbit due to fluctuation in speed at different distances. Finally, the period (T) of a planet's motion squared is inversely proportional to the distance (r) between that planet and the sun cubed.

$$\frac{T^2}{r^3}$$

Sir Isaac Newton used this third law and applied it to the idea of forces and their effects on objects. The effect of the gravitational forces of the Moon on the Earth are noted in the tides, and the effect of the forces of the Earth on the Moon are noted in the fact that the Moon is caught in an orbit around the earth. Since the Moon is traveling at a velocity tangent to its orbit around the Earth and the Earth keeps attracting it in, the Moon does not escape and does not crash into the Earth. The Moon will continue this course due to the attractive gravitational force between the Earth and the Moon. Albert Einstein later applied Newton's adaptation of Kepler's laws. Einstein was able to develop a more advanced theory, which could explain the motions of all the planets and even be applied beyond the solar system. These theories have also been beneficial for predicting behaviors of other objects in the Earth's atmosphere, such as shuttles and astronauts.

Waves and Sound

Mechanical waves are a type of wave that pass through a medium (solid, liquid, or gas). There are two basic types of mechanical waves: longitudinal and transverse.

A **longitudinal wave** has motion that is parallel to the direction of the wave's travel. It can best be shown by compressing one side of a tethered spring and then releasing that end. The movement travels in a bunching and then unbunching motion, across the length of the spring and back until the energy is dissipated through noise and heat.

A **transverse wave** has motion that is perpendicular to the direction of the wave's travel. The particles on a transverse wave do not move across the length of the wave but oscillate up and down to create the peaks and troughs observed on this type of wave.

A wave with a mix of both longitudinal and transverse motion can be seen through the motion of a wave on the ocean, with peaks and troughs, oscillating particles up and down.

Mechanical waves can carry energy, sound, and light. Mechanical waves need a medium through which transport can take place. However, an electromagnetic wave can transmit energy without a medium, or in a vacuum.

Sound travels in waves and is the movement of vibrations through a medium. It can travel through air (gas), land, water, etc. For example, the noise a human hears in the air is the vibration of the waves as they reach the ear. The human brain translates the different frequencies (pitches) and intensities of the vibrations to determine what created the noise.

A tuning fork has a predetermined frequency because of the size (length and thickness) of its tines. When struck, it allows vibrations between the two tines to move the air at a specific rate. This creates a specific tone (or note) for that size of tuning fork. The number of vibrations over time is also steady for that tuning fork and can be matched with a frequency (the number of occurrences over time). All sounds heard by the human ear are categorized by using frequency and measured in **hertz** (the number of cycles per second).

The intensity (or loudness) of sound is measured on the Bel scale. This scale is a ratio of one sound's intensity with respect to a standard value. It is a logarithmic scale, meaning it is measured by factors of ten. But the value that is 1/10 of this value, the decibel, is the measurement used more commonly for the intensity of pitches heard by the human ear.

The Doppler effect applies to situations with both light and sound waves. The premise of the Doppler effect is that, based on the relative position or movement of a source and an observer, waves can seem shorter or longer than they are. When the Doppler effect is experienced with sound, it warps the noise being heard by the observer by making the pitch or frequency seem shorter or higher as the source is approaching and then longer or lower as the source is getting farther away. The frequency and pitch of the source never actually change, but the sound in respect to the observer's position makes it seem like the sound has changed. This effect can be observed when an emergency siren passes by an observer on the road. The siren sounds much higher in pitch as it approaches the observer and then lower after it passes and is getting farther away.

The Doppler effect also applies to situations involving light waves. An observer in space would see light approaching as being shorter wavelengths than it was, causing it to appear blue. When the light wave is getting farther away, the light would look red due to the apparent elongation of the wavelength. This is called the **red-blue shift**.

A recent addition to the study of waves is the gravitational wave. Its existence has been proven and verified, yet the details surrounding its capabilities are still under inquiry. Further understanding of gravitational waves could help scientists understand the beginnings of the universe and how the existence of the solar system is possible. This understanding could also include the future exploration of the universe.

Light

The movement of light is described like the movement of waves. Light travels with a wave front and has an amplitude (a height measured from the neutral), a cycle or wavelength, a period, and energy. Light travels at approximately 3.00×10^8 m/s and is faster than anything created by humans.

Light is commonly referred to by its measured **wavelengths**, or the length for it to complete one cycle. Types of light with the longest wavelengths include radio, TV, micro, and infrared waves. The next set of wavelengths are detectable by the human eye and make up the visible spectrum. The visible spectrum has wavelengths of 10^{-7} m, and the colors seen are red, orange, yellow, green, blue, indigo, and violet. Beyond the visible spectrum are even shorter wavelengths (also called the **electromagnetic spectrum**) containing ultraviolet light, x-rays, and gamma rays. The wavelengths outside of the visible light range can be harmful to humans if they are directly exposed, especially for long periods of time.

For example, the light from the sun has a small percentage of ultraviolet (UV) light, which is mostly absorbed by the UV layer of the Earth's atmosphere. When this layer does not filter out the UV rays, the exposure to the wavelengths can be harmful to human skin. When there is an extra layer of pollutant and the light from the sun is trapped by repeated reflection back to the ground, unable to bounce back into space, it creates another harmful condition for the planet called the **greenhouse effect**. This is an overexposure to the sun's light and contributes to global warming by increasing the temperatures on Earth.

When a wave crosses a boundary or travels from one medium to another, certain actions take place. If the wave travels through one medium into another, it experiences **refraction**, which is the bending of the wave from one medium's density to another, altering the speed of the wave.

For example, a side view of a pencil in half a glass of water appears as though it is bent at the water level. What the viewer is seeing is the refraction of light waves traveling from the air into the water. Since the wave speed is slowed in water, the change makes the pencil appear bent.

When a wave hits a medium that it cannot pass through, it is bounced back in an action called **reflection**. For example, when light waves hit a mirror, they are reflected, or bounced off, the back of the mirror. This can cause it to seem like there is more light in the room due to the doubling back of the initial wave. This is also how people can see their reflection in a mirror.

When a wave travels through a slit or around an obstacle, it is known as **diffraction**. A light wave will bend around an obstacle or through a slit and cause a diffraction pattern. When the waves bend around an obstacle, it causes the addition of waves and the spreading of light on the other side of the opening.

Optics

The dispersion of light describes the splitting of a single wave by refracting its components into separate parts. For example, if a wave of white light is sent through a dispersion prism, the light wave appears as its separate rainbow-colored components due to each colored wavelength being refracted in the prism.

Different things occur when wavelengths of light hit boundaries. Objects can absorb certain wavelengths of light and reflect others, depending on the boundaries. This becomes important when an object appears to be a certain color. The color of the object is not actually within the makeup of that object, but by what wavelengths are being transmitted by that object. For example, if a table appears to be red, that means the table is absorbing all wavelengths of visible light except those of the red wavelength. The table is reflecting, or transmitting, the wavelengths associated with red back to the human eye, and therefore, the table appears red.

Interference describes when an object affects the path of a wave or another wave interacts with that wave. Waves interacting with each other can result in either constructive interference or destructive interference based on their positions. For constructive interference, the waves are in sync and combine to reinforce each other. In the case of deconstructive interference, the waves are out of sync and reduce the effect of each other to some degree. In scattering, the boundary can change the direction or energy of a wave, thus altering the entire wave. Polarization changes the oscillations of a wave and can alter its appearance in light waves. For example, polarized sunglasses take away the "glare" from sunlight by altering the oscillation pattern observed by the wearer.

When a wave hits a boundary and is completely reflected or cannot escape from one medium to another, it is called **total internal reflection**. This effect can be seen in a diamond with a brilliant cut. The angle cut on the sides of the diamond causes the light hitting the diamond to be completely reflected inside the gem and makes it appear brighter and more colorful than a diamond with different angles cut into its surface.

When reflecting light, a mirror can be used to observe a virtual (not real) image. A plane mirror is a piece of glass with a coating in the background to create a reflective surface. An image is what the human eye sees when light is reflected off the mirror in an unmagnified manner. If a curved mirror is used for reflection, the image seen will not be a true reflection, but will either be magnified or made to appear smaller than its actual size. Curved mirrors can also make an object appear closer or farther away than its actual distance from the mirror.

Lenses can be used to refract or bend light to form images. Examples of lenses are human eye, microscopes, and telescopes. The human eye interprets the refraction of light into images that humans understand to be actual size. When objects are too small to be observed by the unaided human eye, microscopes allow the objects to be enlarged enough to be seen. Telescopes allow objects that are too far away to be seen by the unaided eye to be viewed. Prisms are pieces of glass that can have a wavelength of light enter one side and appear to be broken down into its component wavelengths on the other side, due to the slowing of certain wavelengths within the prism, more than other wavelengths.

Nature of Electricity

Electrostatics is the study of electric charges at rest. A balanced atom has a neutral charge from its number of electrons and protons. If the charge from its electrons is greater than or less than the charge of its protons, the atom has a charge. If the atom has a greater charge from the number of electrons than protons, it has a negative charge. If the atom has a lesser charge from the number of electrons than protons, it has a positive charge. Opposite charges attract each other, while like charges repel each other, so a negative attracts a positive, and a negative repels a negative. Similarly, a positive charge repels a positive charge. Just as energy cannot be created or destroyed, neither can charge; charge can only be transferred. The transfer of charge can occur through touch, or the transfer of electrons. Once electrons have transferred from one object to another, the charge has been transferred.

For example, if a person wears socks and scuffs their feet across a carpeted floor, the person is transferring electrons to the carpeting through the friction from their feet. Additionally, if that person then touches a light switch, they receive a small shock. This "shock" is the person feeling the electrons transferring from the switch to their hand. Since the person lost electrons to the carpet, that person now has fewer negative charges, resulting in a net positive charge. Therefore, the electrons from the light switch are attracted to the person for the transfer. The shock felt is the electrons moving from the switch to the person's finger.

Another method of charging an object is through induction. **Induction** occurs when a charged object is brought near two touching stationary objects. The electrons in the objects will attract and cluster near another positively-charged object and repel away from a negatively-charged object held nearby. The stationary objects will redistribute their electrons to allow the charges to reposition themselves closer or farther away. This redistribution will cause one of the touching stationary objects to be negatively charged and the other to be positively charged. The overall charges contained in the stationary objects remain the same but are repositioned between the two objects.

Another way to charge an object is through **polarization**. Polarization can occur simply by the reconfiguration of the electrons within a single object.

For example, if a girl at a birthday party rubs a balloon on her hair, the balloon could then cling to a wall if it were brought close enough. This would be because rubbing the balloon causes it to become negatively charged. When the balloon is held against a neutrally-charged wall, the negatively charged balloon repels all the wall's electrons, causing a positively-charged surface on the wall. This type of charge is temporary, due to the massive size of the wall, and the charges will quickly redistribute.

An electric current is produced when electrons carry charge across a length. To make electrons move so they can carry this charge, a change in voltage must be present. On a small scale, this is demonstrated through the electrons traveling from the light switch to a person's finger in the example where the person had run their socks on a carpet. The difference between the charge in the switch and the charge in the finger causes the electrons to move. On a larger and more sustained scale, this movement would need to be more controlled. This can be achieved through batteries/cells and generators. Batteries or cells have a chemical reaction that takes place inside, causing energy to be released and charges to move freely. Generators convert mechanical energy into electric energy for use after the reaction.

For example, if a wire runs touching the end of a battery to the end of a lightbulb, and then another wire runs touching the base of the lightbulb to the opposite end of the original battery, the lightbulb will light up. This is due to a complete circuit being formed with the battery and the electrons being carried across the voltage drop (the two ends of the battery). The appearance of the light from the bulb is the visible presence of the electrons in the filament of the bulb.

Electric energy can be derived from a number of sources, including coal, wind, sun, and nuclear reactions. Electricity has numerous applications, including being transferable into light, sound, heat, or magnetic forces.

Magnetism and Electricity

Magnetic forces occur naturally in specific types of materials and can be imparted to other types of materials. If two straight iron rods are observed, they will naturally have a negative end (pole) and a positive end (pole). These charged poles follow the rules of any charged item: Opposite charges attract, and like charges repel. When set up positive to negative, they will attract each other, but if one rod is turned around, the two rods will repel each other due to the alignment of negative to negative poles and positive to positive poles. When poles are identified, magnetic fields are observed between them.

If small iron filings (a material with natural magnetic properties) are sprinkled over a sheet of paper resting on top of a bar magnet, the field lines from the poles can be seen in the alignment of the iron filings, as pictured below:

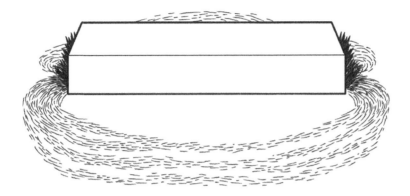

These fields naturally occur in materials with magnetic properties. There is a distinct pole at each end of such a material. If materials are not shaped with definitive ends, the fields will still be observed through the alignment of poles in the material. For example, a circular magnet does not have ends but still has a magnetic field associated with its shape, as pictured below:

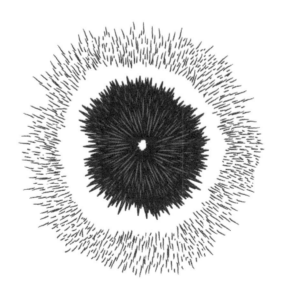

Magnetic forces can also be generated and amplified by using an electric current. For example, if an electric current is sent through a length of wire, it creates an electromagnetic field around the wire from the charge of the current. This force is from the moving of negatively-charged electrons from one end of the wire to the other. This is maintained as long as the flow of electricity is sustained. The magnetic field can also be used to attract and repel other items with magnetic properties. A smaller or larger magnetic force can be generated around this wire, depending on the strength of the current in the wire. As soon as the current is stopped, the magnetic force also stops.

Magnetic energy can be harnessed, or manipulated, from natural sources or from a generated source (a wire carrying electric current). When a core with magnetic properties (such as iron) has a wire wrapped around it in circular coils, it can be used to create a strong, non-permanent electromagnet. If current is run through the wrapped wire, it generates a magnetic field by polarizing the ends of the metal core, as described above, by moving the negative charge from one end to the other. If the direction of the current is reversed, so is the direction of the magnetic field due to the poles of the core being reversed. The term **non-permanent** refers to the fact that the magnetic field is generated only when the current is present, but not when the current is stopped. The following is a picture of a small electromagnet made from an iron nail, a wire, and a battery:

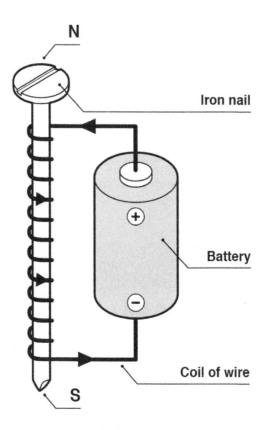

This type of electromagnetic field can be generated on a larger scale using more sizable components. This type of device is useful in the way it can be controlled. Rather than having to attempt to block a permanent magnetic field, the current to the system can simply be stopped, thus stopping the magnetic field. This provides the basis for many computer-related instruments and magnetic resonance imaging (MRI) technology. Magnetic forces are used in many modern applications, including the creation of super-speed transportation. Super magnets are used in rail systems and supply a cleaner form of energy than coal or gasoline.

Another example of the use of super-magnets is seen in medical equipment, specifically MRI. These machines are highly sophisticated and useful in imaging the internal workings of the human body. For super-magnets to be useful, they often must be cooled down to extremely low temperatures to dissipate the amount of heat generated from their extended usage. This can be done by flooding the magnet with a super-cooled gas such as helium or liquid nitrogen. Much research is continuously done in this field to find new ceramic–metallic hybrid materials that have structures that can maintain their charge and temperature within specific guidelines for extended use.

Biology

Biology Basics

Atoms are the smallest units of all matter and make up all chemical elements. They each have three parts: protons, neutrons, and electrons. **Protons** are found in the nucleus of an atom and have a positive electric charge. They have a mass of about one atomic mass unit. The number of protons in an element is referred to as the element's **atomic number**. Each element has a unique number of protons, and therefore a unique atomic number. **Neutrons** are also found in the nucleus of atoms. These subatomic particles have a neutral charge, meaning that they do not have a positive or negative electric charge. Their mass is slightly larger than that of a proton.

Together with protons, they are referred to as the **nucleons** of an atom. The **atomic mass** number of an atom is equal to the sum of the protons and neutrons in the atom. **Electrons** have a negative charge and are the smallest of the subatomic particles. They are located outside the nucleus in **orbitals**, which are shells that surround the nucleus. If an atom has an overall neutral charge, it has an equal number of electrons and protons. If it has more protons than electrons or vice versa, it becomes an **ion**. When there are more protons than electrons, the atom is a positively-charged ion, or **cation**. When there are more electrons than protons, the atom is a negatively-charged ion, or **anion**.

The location of electrons within an atom is more complicated than the locations of protons and neutrons. Within the orbitals, electrons are always moving. They can spin very fast and move upward, downward, and sideways. There are many different levels of orbitals around the atomic nucleus, and each orbital has a different capacity for electrons. The electrons in the orbitals closest to the nucleus are more tightly bound to the nucleus of the atom. There are three main characteristics that describe each orbital. The first is the **principal quantum number**, which describes the size of the orbital. The second is the **angular momentum quantum number**, which describes the shape of the orbital. The third is the **magnetic quantum number**, which describes the orientation of the orbital in space.

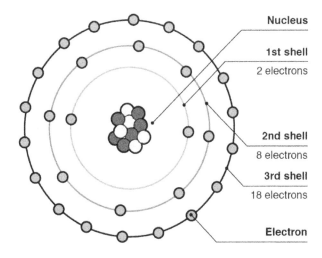

Another important characteristic of electrons is their ability to form covalent bonds with other atoms to form molecules. A **covalent bond** is a chemical bond that forms when two atoms share the same pair or pairs of electrons. There is a stable balance of attraction and repulsion between the two atoms. There are several types of covalent bonds. **Sigma bonds** are the strongest type of covalent bond and involve the head-on overlapping of electron orbitals from two different atoms. **Pi bonds** are a little weaker and involve the lateral overlapping of certain orbitals. While single bonds between atoms, such as between carbon and hydrogen, are generally sigma bonds, double bonds, such as when carbon is double-bonded to an oxygen atom, are usually formed from one sigma bond and one pi bond.

The Cell

All living organisms are made up cells. **Cells** are considered the basic functional unit of organisms and the smallest unit of matter that is living. Most organisms are multicellular, which means that they are made up of more than one cell and often they are made up of a variety of different types of cells. Cells contain **organelles**, which are the little working parts of the cell, responsible for specific functions that keep the cell and organism alive.

Plant and animal cells have many of the same organelles but also have some unique traits that distinguish them from each other. Plants contain a cell wall, while animal cells are only surrounded by a phospholipid plasma membrane. The cell wall is made up of strong, fibrous polysaccharides and proteins. It protects the cell from mechanical damage and maintains the cell's shape. Inside the cell wall, plant cells also have plasma membrane. The plasma membrane of both plant and animal cells is made up of two layers of phospholipids, which have a hydrophilic head and hydrophobic tails. The tails converge towards each other on the inside of the bilayer, while the heads face the interior of the cell and the exterior environment. **Microvilli** are protrusions of the cell membrane that are only found in animal cells. They increase the surface area and aid in absorption, secretion, and cellular adhesion. Chloroplasts are also only found in plant cells. They are responsible for photosynthesis, which is how plants convert sunlight into chemical energy.

The list below describes major organelles that are found in both plant and animal cells:

- **Nucleus**: The nucleus contains the DNA of the cell, which has all the cells' hereditary information passed down from parent cells. DNA and protein are wrapped together into chromatin within the nucleus. The nucleus is surrounded by a double membrane called the nuclear envelope.

- **Endoplasmic Reticulum (ER)**: The ER is a network of tubules and membranous sacs that are responsible for the metabolic and synthetic activities of the cell, including synthesis of membranes. Rough ER has ribosomes attached to it, while smooth ER does not.

- **Mitochondrion**: The mitochondrion is essential for maintaining regular cell function and is known as the powerhouse of the cell. It is where cellular respiration occurs and where most of the cell's ATP is generated.

- **Golgi Apparatus**: The Golgi apparatus is where cell products are synthesized, modified, sorted, and secreted out of the cell.

- **Ribosomes**: Ribosomes make up a complex that produces proteins within the cell. They can be free in the cytosol or bound to the ER.

Cellular Respiration

Cellular respiration in multicellular organisms occurs in the mitochondria. It is a set of reactions that converts energy from nutrients to ATP and can either use oxygen in the process, which is called **aerobic respiration**, or not, which is called **anaerobic respiration**.

Aerobic respiration has two main parts, which are the citric acid cycle, also known as Krebs cycle, and oxidative phosphorylation. Glucose is a commonly-found molecule that is used for energy production within the cell. Before the citric acid cycle can begin, the process of glycolysis converts glucose into two pyruvate molecules. Pyruvate enters the mitochondrion, is oxidized, and then is converted to a compound called acetyl CoA. There are eight steps in the citric acid cycle that start with acetyl CoA and convert it to oxaloacetate and NADH. The oxaloacetate continues in the citric acid cycle and the NADH molecule moves on to the oxidative phosphorylation part of cellular respiration. Oxidative phosphorylation has two main steps, which are the electron transport chain and chemiosmosis. The mitochondrial membrane has four protein complexes within it that help to transport electrons through the inner mitochondrial matrix.

Electrons and protons are removed from NADH and $FADH_2$ and then transported along these and other membrane complexes. Protons are pumped across the inner membrane, which creates a gradient to draw electrons to the intermembrane complexes. Two mobile electron carriers, ubiquinone and cytochrome C, are also located in the inner mitochondrial membrane. At the end of these electron transport chains, the electrons are accepted by O_2 molecules and water is formed with the addition of two hydrogen atoms. Chemiosmosis occurs in an ATP synthase complex that is located next to the four electron transport complexes. As the complex pumps protons from the intermembrane space to the mitochondrial matrix, ADP molecules become phosphorylated and ATP molecules are generated. Approximately four to six ATP molecules are generated during glycolysis and the citric acid cycle and twenty-six to twenty-eight ATP molecules are generated during oxidative phosphorylation, which makes the total number of ATP molecules generated during aerobic cellular respiration approximately thirty to thirty-two.

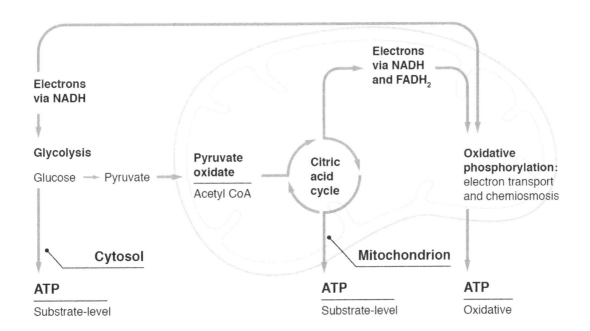

Since not all environments are oxygen-rich, some organisms must find alternate ways to extract energy from nutrients. The process of anaerobic respiration is similar to that of aerobic respiration in that protons and electrons are removed from nutrient molecules and are passed down an electron transport chain, with the end result being ATP synthesis. However, instead of the electrons being accepted by oxygen molecules at the end of the electron transport chain, they are accepted by either sulfate or nitrate molecules. Anaerobic respiration is mostly used by unicellular organisms, or prokaryotic organisms.

Photosynthesis

Photosynthesis is a set of reactions that occur to convert light energy into chemical energy. The chemical energy is then stored as sugar and other organic molecules inside the organism or plant. Within plants, the photosynthetic process takes place within chloroplasts. The two stages of photosynthesis are the light reactions and the Calvin cycle. Within chloroplasts, there are membranous sacs called **thylakoids** and within the thylakoids is the green pigment called **chlorophyll**. The light reactions take place in the chlorophyll. The Calvin cycle takes place in the **stroma**, or inner space, or the chloroplasts.

During the light reactions, light energy is absorbed by chlorophyll. First, a light-harvesting complex, called photosystem II (PS II), absorbs photons from light that enters the chlorophyll and then passes it onto a reaction-center complex. Once the photon enters the reaction-center complex, it causes a special pair of chlorophyll *a* molecules to release an electron. The electron is accepted by a primary electron acceptor molecule, while at the same time, a water molecule is dissociated into two hydrogen atoms, one oxygen atom, and two electrons. These electrons are transferred to the chlorophyll *a* molecules that just lost their electrons.

The electrons that were released from the chlorophyll *a* molecules move down an electron transport chain using an electron carrier, called plastoquinone, a cytochrome complex, and a protein, called plastocyanin. At the end of the chain, the electrons reach another light-harvesting complex, called photosystem I (PS I). While in the cytochrome complex, the electrons cause protons to be pumped into the thylakoid space, which in turn provides energy for ATP molecules to be produced. A primary electron acceptor molecule accepts the electrons that are released from PS I and then passes them onto another electron transport chain, which includes the protein ferredoxin. At the end of the light reactions, electrons are transferred from ferredoxin to NADP+, producing NADPH. The ATP and NADPH that are produced through the light reactions are used as energy to drive the Calvin cycle forward.

The three phases of the Calvin cycle are carbon fixation, reduction, and regeneration of the CO_2 acceptor. Carbon fixation occurs when CO_2 is introduced into the cycle and attaches to a five-carbon sugar, called ribulose bisphosphate (RuBP). A six-carbon sugar is split into two three-carbon sugar molecules, known as 3-phosphoglycerate. Next, during the reduction phase, an ATP molecule loses a phosphate group and becomes ADP. The phosphate group attaches to the 3-phosphoglycerate molecule, making it 1,3-bisphosphate. Then, an NADPH molecule donates two electrons to this new molecule, causing it to lose a phosphate group and become glyceraldehyde 3-phosphate (G3P), a sugar molecule. At the end of the cycle, one G3P molecule exits the cycle and is used by the plant for energy. Five other G3P molecules continue in the cycle to regenerate RuBP molecules, which are the CO_2 acceptors of the cycle. When every photon has been used up, three RuBP molecules are formed from the rearrangement of five G3P molecules and wait for the cycle to start again. It takes three turns of the cycle and three CO_2 molecules entering the cycle to generate just one G3P molecule.

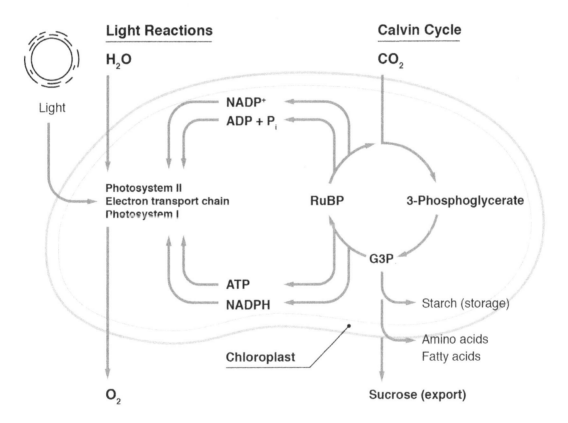

Cellular Reproduction

Cellular reproduction is the process that cells follow to make new cells with the same or similar contents as themselves. This process is an essential part of an organism's life. It allows for the organism to grow larger itself, and as it ages, it allows for replacement of dying and damaged cells. The process of cellular reproduction must be accurate and precise. Otherwise, the new cells that are produced will not be able to perform the same functions as the original cell. Mutations can occur in the offspring, which can cause anywhere from minor to severe problems.

The two types of cellular reproduction that organisms can use are mitosis or meiosis. **Mitosis** produces daughter cells that are identical to the parent cell and is often referred to as **asexual reproduction**. **Meiosis** has two stages of cell division and produces daughter cells that have a combination of traits from their two parents. It is often referred to as **sexual reproduction**. Humans reproduce by meiosis. During this process, the sperm, the male germ cell, and the egg, the female germ cell, combine and form a parent cell that contains both of their sets of chromosomes. This parent cell then divides into four daughter cells, each with a unique set of traits that came from both of their parents.

In both processes, the most important part of the cell that is copied is the cell's DNA. It contains all of the genetic information for the cell, which leads to its traits and capabilities. Some parts of the cell are copied exactly during cellular reproduction, such as DNA. However, certain other cellular components are synthesized within the new cell after reproduction is complete using the new DNA. For example, the endoplasmic reticulum is broken down during the cell cycle and then newly synthesized after cell division.

Genetics

Humans carry their genetic information on structures called **chromosomes**. Chromosomes are string-like structures made up of nucleic acids and proteins. During the process of reproduction, each parent contributes a gamete that contains twenty-three chromosomes to the intermediate diploid cell. This diploid cell, which contains forty-six chromosomes, replicates itself to produce two diploid cells. These two diploid cells each then split into cells and randomly divide the chromosomes so that each of the four resulting cells contains only twenty-three chromosomes, as each parent gamete does.

Each of the twenty-three chromosomes has between a few hundred and a few thousand genes on it. Each gene contains information about a specific trait that is inherited by the offspring from one of the parents. **Genes** are made up of sequences of DNA that encode proteins and start pathways to express the phenotype that they control. Every gene has two **alleles**, or variations—one inherited from each parent. In most genes, one allele has a more dominant phenotype than the other allele. This means that when both alleles are present on a gene, the dominant phenotype will always be expressed over the recessive phenotype.

The recessive phenotype would only be expressed when both alleles present were recessive. Some alleles can have codominance or incomplete dominance. **Codominance** occurs when both alleles are expressed equally when they are both present. For example, the hair color of cows can be red or white, but when one allele for each hair color is present, their hair is a mix of red and white, not pink. **Incomplete dominance** occurs when the presence of two different alleles creates a third phenotype. For example, some flowers can be red or white when the alleles are in duplicate, but when one of each allele is present, the flowers are pink.

Mendel's Laws of Heredity

Gregor Mendel was a monk who came up with one of the first models of inheritance in the 1860s. He is often referred to as the father of genetics. At the time, his theories were largely criticized because biologists did not believe that these ideas could be generally applicable and could also not apply to different species. They were later rediscovered in the early 1900s and given more credence by a group of European scientists. Mendel's original ideas have since been combined with other theories of inheritance to develop the ideas that are studied today.

Between 1856 and 1863, Gregor Mendel experimented with about five thousand pea plants in his garden that had different color flowers to test his theories of inheritance. He crossed purebred white flower and purple flower pea plants and found that the results were not a blend of the two flowers; they were instead all purple flowers. When he then fertilized this second generation of purple flowers with itself, both white flowers and purple flowers were produced, in a ratio of one to three. Although he used different terms at the time, he proposed that the color trait for the flowers was regulated by a gene, which he called a

46

"factor," and that there were two alleles, which he called "forms," for each gene. For each gene, one allele was inherited from each parent. The results of these experiments allowed him to come up with his Principles of Heredity.

There are two main laws that Mendel developed after seeing the results of his experiments. The first law is the **Law of Segregation**, which states that each trait has two versions that can be inherited. In the parent cells, the allele pairs of each gene separate, or segregate, randomly during gamete production. Each gamete then carries only one allele with it for reproduction. During the process of reproduction, the gamete from each parent contributes its single allele to the daughter cell. The second law is the **Law of Independent Assortment**, which states that the alleles for different traits are not linked to one another and are inherited independently. It emphasizes that if a daughter cell selects allele A for gene 1, it does not also automatically select allele A for gene 2. The allele for gene 2 is selected in a separate, random manner.

Mendel theorized one more law, called the **Law of Dominance**, which has to do with the expression of a genotype but not with the inheritance of a trait. When he crossed the purple flower and white flower pea plants, he realized that the purple flowers were expressed at a greater ratio than the white flower pea plants. He hypothesized that certain gene alleles had a stronger outcome on the phenotype that was expressed. If a gene had two of the same allele, the phenotype associated with that allele was expressed. If a gene had two different alleles, the phenotype was determined by the dominant allele, and the other allele, the recessive allele, had no effect on the phenotype.

DNA

DNA is made up of two polynucleotide strands that are linked together and twisted into a double-helix structure. The polynucleotide strands are made up a chain of nucleotides made from four nitrogenous bases, which are adenine, thymine, guanine, and cytosine. Across the two strands, adenine and thymine are paired together, and guanine and cytosine are paired together. This allows for a tight helix to form based on their molecular configurations. DNA contains all the genetic information of a living organism and provides the protein encoding information on genes.

Chromosome replication and cell division start with DNA replication. It is the first step in passing genetic information to subsequent generations. During replication, the double-helix DNA is untwisted and separated. Each strand is replicated and then linked to one of the original strands to form two new DNA molecules. Sometimes, there are errors made in the DNA sequence that codes for a specific gene, causing genetic mutations. If the sequence alteration was originally in the parent gene, it is considered hereditary. If it was developed during the replication process, it is classified as an acquired mutation. Some mutations can cause major phenotypic differences, resulting in developmental problems, while others do not affect development at all. Although they are not very common, mutations are important for the variation of the general population.

Practice Quiz

1. In what part of the female reproductive system do sperm fertilize an oocyte?
   a. Ovary
   b. Mammary gland
   c. Vagina
   d. Fallopian tubes

2. What is the largest organ in the human body?
   a. Brain
   b. Large intestine
   c. Skin
   d. Liver

3. Which of the following is a function of the skin?
   a. Temperature regulation
   b. Breathing
   c. Ingestion
   d. Gas exchange

4. Which accessory organ of the integumentary system provides a hard layer of protection over the skin?
   a. Hair
   b. Nails
   c. Sweat glands
   d. Sebaceous glands

5. Which two magnetic poles would attract each other?
   a. Two positive poles
   b. Two negative poles
   c. One positive pole and one negative pole
   d. Two circular positive magnets

See answers on next page.

Answer Explanations

1. D: Once sperm enter the vagina, they travel through the uterus to the Fallopian tubes to fertilize a mature oocyte. The ovaries are responsible for producing the mature oocyte. The mammary glands produce nutrient-filled milk to nourish babies after birth.

2. C: The skin is the largest organ of the body, covering every external surface of the body and protecting the body's deeper tissues. The brain performs many complex functions, and the large intestine is part of the gastrointestinal system, but neither of these is largest organ in the body. The liver is the second largest organ of the body, weighing roughly three pounds in the average adult.

3. A: The skin has three major functions: protection, regulation, and sensation. It has a large supply of blood vessels that can dilate to allow heat loss when the body is too hot and constrict in order to retain heat when the body is cold. The organs of the respiratory system are responsible for breathing and work together with the circulatory system for gas exchange, and the mouth is responsible for ingestion.

4. B: The nails on a person's hands and feet provide a hard layer of protection over the soft skin underneath. The hair, Choice *A*, helps to protect against heat loss, provides sensation, and filters air that enters the nose. The sweat glands, Choice *C*, help to regulate temperature. The sebaceous glands, Choice *D*, secrete sebum, which protects the skin from water loss and bacterial and fungal infections.

5. C: Magnets follow the same rules of charge as other items. Positive and negative poles attract each other. Two poles with the same charge, positive or negative, would repel each other; therefore, Choices *A* and *B* are incorrect. Although circular magnets do not have ends, they still have poles and follow the same rules of charge. So, circular positive magnets would repel each other, making Choice *D* incorrect.

Arithmetic Reasoning

Numbers and Algebra

Definitions

Whole numbers are the numbers 0, 1, 2, 3, ... Examples of other whole numbers would be 413 and 8,431. Notice that numbers such as 4.13 and $\frac{1}{4}$ are not included in whole numbers. **Counting numbers**, also known as **natural numbers**, consist of all whole numbers except for the zero. In set notation, the natural numbers are the set $\{1, 2, 3, ... \}$. The entire set of whole numbers and negative versions of those same numbers comprise the set of numbers known as **integers**. Therefore, in set notation, the integers are $\{..., -3, -2, -1, 0, 1, 2, 3, ... \}$. Examples of other integers are −4,981 and 90,131. A number line is a great way to visualize the integers. Integers are labeled on the following number line:

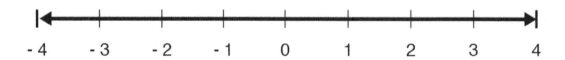

The arrows on the right- and left-hand sides of the number line show that the line continues indefinitely in both directions.

Fractions also exist on the number line as parts of a whole. For example, if an entire pie is cut into two pieces, each piece is half of the pie, or $\frac{1}{2}$. The top number in any fraction, known as the **numerator**, defines how many parts there are. The bottom number, known as the **denominator**, states how many pieces the whole is divided into. Fractions can also be negative or written in their corresponding decimal form.

A **decimal** is a number that uses a decimal point and numbers to the right of the decimal point representing the part of the number that is less than 1. For example, 3.5 is a decimal and is equivalent to the fraction $\frac{7}{2}$ or the mixed number $3\frac{1}{2}$. The decimal is found by dividing 2 into 7. Other examples of fractions are $\frac{2}{7}, \frac{-3}{14}$, and $\frac{14}{27}$.

Any number that can be expressed as a fraction is known as a **rational number**. Basically, if a and b are any integers and $b \neq 0$, then $\frac{a}{b}$ is a rational number. Any integer can be written as a fraction where the denominator is 1, so therefore the rational numbers consist of all fractions and all integers.

Any number that is not rational is known as an **irrational number**. Consider the number:

$$\pi = 3.141592654 ...$$

The decimal portion of that number extends indefinitely. In that situation, a number can never be written as a fraction. Another example of an irrational number is:

$$\sqrt{2} = 1.414213662 ...$$

Again, this number cannot be written as a ratio of two integers.

Together, the set of all rational and irrational numbers makes up the real numbers. The number line contains all real numbers. To graph a number other than an integer on a number line, it needs to be plotted between two integers. For example, 3.5 would be plotted halfway between 3 and 4.

Even numbers are integers that are divisible by 2. For example, 6, 100, 0, and −200 are all even numbers. **Odd numbers** are integers that are not divisible by 2. If an odd number is divided by 2, the result is a fraction. For example, −5, 11, and −121 are odd numbers.

Prime numbers consist of natural numbers greater than 1 that are not divisible by any other natural numbers other than themselves and 1. For example, 3, 5, and 7 are prime numbers. If a natural number is not prime, it is known as a **composite number**. 8 is a composite number because it is divisible by both 2 and 4, which are natural numbers other than itself and 1.

The **absolute value** of any real number is the distance from that number to 0 on the number line. The absolute value of a number can never be negative. For example, the absolute value of both 8 and −8 is 8 because they are both 8 units away from 0 on the number line. This is written as:

$$|8| = |-8| = 8$$

Writing Numbers Using Base-10 Numerals, Number Names, and Expanded Form

The **base-10 number system** is also called the **decimal system of naming numbers**. There is a decimal point that sets the value of numbers based on their position relative to the decimal point. The order from the decimal point to the right is the tenths place, then hundredths place, then thousandths place. Moving to the left from the decimal point, the place value is ones, tens, hundreds, etc. The number 2,356 can be described in words as "two thousand three hundred fifty-six." In expanded form, it can be written as:

$$(2 \times 1,000) + (3 \times 100) + (5 \times 10) + (6 \times 1)$$

The expanded form shows the value each number holds in its place. The number 3,093 can be written in words as "three thousand ninety-three." In expanded form, it can be expressed as:

$$(3 \times 1,000) + (0 \times 100) + (9 \times 10) + (3 \times 1)$$

Notice that the zero is added in the expanded form as a place holder. There are no hundreds in the number, so a zero is written in the hundreds place.

Composing and Decomposing Multidigit Numbers

Composing and decomposing numbers reveals the place value held by each number 0 through 9 in each position. For example, the number 17 is read as "seventeen." It can be decomposed into the numbers 10 and 7. It can be described as 1 group of ten and 7 ones. The one in the tens place represents one set of ten. The seven in the ones place represents seven sets of one. Added together, they make a total of seventeen. The number 48 can be written in words as "forty-eight." It can be decomposed into the numbers 40 and 8, where there are 4 groups of ten and 8 groups of one. The number 296 can be decomposed into 2 groups of one hundred, 9 groups of ten, and 6 groups of one. There are two hundreds, nine tens, and six ones. Decomposing and composing numbers lays the foundation for visually picturing the number and its place value, and adding and subtracting multiple numbers with ease.

Place and Value of a Digit

Each number in the base-10 system is made of the numbers 0–9, located in different places relative to the decimal point. Based on where the numbers fall, the value of a digit changes. For example, the number 7,509 has a seven in the thousands place. This means there are seven groups of one thousand. The number 457 has a seven in the ones place. This means there are seven groups of one. Even though there is a seven in both numbers, the place of the seven tells the value of the digit. A practice question may ask the place and value of the 4 in 3,948. The four is found in the tens place, which means four represents the number 40, or four groups of ten. Another place value may be on the opposite side of the decimal point. A question may ask the place and value of the 8 in the number 203.80. In this case, the eight is in the tenths place because it is in the first place to the right of the decimal point. It holds a value of eight-tenths, or eight groups of one-tenth.

Relative Value of a Digit

The value of a digit is found by recognizing its place relative to the rest of the number. For example, the number 569.23 contains a 6. The position of the 6 is two places to the left of the decimal, putting it in the tens place. The tens place gives it a value of 60, or six groups of ten. The number 39.674 has a 4 in it. The number 4 is located three places to the right of the decimal point, placing it in the thousandths place. The value of the 4 is four-thousandths, because of its position relative to the other numbers and to the decimal. It can be described as 0.004 by itself, or four groups of one-thousandths. The numbers 100 and 0.1 are both made up of ones and zeros. The first number, 100, has a 1 in the hundreds place, giving it a value of one hundred. The second number, 0.1, has a 1 in the tenths place, giving that 1 a value of one-tenth. The place of the number gives it the value.

Rounding Multidigit Numbers

Numbers can be rounded by recognizing the place value where the rounding takes place, then looking at the number to the right. If the number to the right is five or greater, the number to be rounded goes up one. If the number to the right is four or less, the number to be rounded stays the same. For example, the number 438 can be rounded to the tens place. The number 3 is in the tens place and the number to the right is 8.

Because the 8 is 5 or greater, the 3 then rounds up to a 4. The rounded number is 440. Another number, 1,394, can be rounded to the thousands place. The number in the thousands place is 1, and the number to the right is 3. As the 3 is 4 or less, it means the 1 stays the same and the rounded number is 1,000. Rounding is also a form of estimating. The number 9.58 can be rounded to the tenths place. The number 5 is in the tenths place, and the number 8 is to the right of it. Because 8 is 5 or greater, the 5 changes to a 6. The rounded number becomes 9.6.

Prime and Composite Numbers

A **prime number** is a whole number greater than 1 that can only be divided by 1 and itself. Examples are 2, 3, 5, 7, and 11. A **composite number** can be evenly divided by a number other than 1 and itself. Examples of composite numbers are 4 and 9. Four can be divided evenly by 1, 2, and 4. Nine can be divided evenly by 1, 3, and 9. When given a list of numbers, one way to determine which ones are prime or composite is to find the **prime factorization** of each number. For example, a list of numbers may include 13 and 24. The prime factorization of 13 is 1 and 13 because those are the only numbers that go into it evenly, so it is a prime number. The prime factorization of 24 is 2 × 2 × 2 × 3 because those are

52

the prime numbers that multiply together to get 24. This also shows that 24 is a composite number because 2 and 3 are factors along with 1 and 24.

Multiples of Numbers

A **multiple** of a number is the result of multiplying that number by an integer. For example, some multiples of 3 are 6, 9, 12, 15, and 18. These multiples are found by multiplying 3 by 2, 3, 4, 5, and 6, respectively. Some multiples of 5 include 5, 10, 15, and 20. This also means that 5 is a factor of 5, 10, 15, and 20. Some questions may ask which numbers in a list are multiples of a given number. For example, find and circle the multiples of 12 in the following list: 136, 144, 312, 400. If a number is evenly divisible by 12, then it is a multiple of 12. The numbers 144 and 312 are multiples of 12 because 12 times 12 is 144, and 12 times 26 is 312. The other numbers, 136 and 400, are not multiples because they yield a number with a fractional component when divided by 12.

Factorization

Factorization is the process of breaking up a mathematical quantity, such as a number or polynomial, into a product of two or more factors. For example, a factorization of the number 16 is:

$$16 = 8 \times 2$$

If multiplied out, the factorization results in the original number. A **prime factorization** is a specific factorization when the number is factored completely using prime numbers only. For example, the prime factorization of 16 is:

$$16 = 2 \times 2 \times 2 \times 2$$

A factor tree can be used to find the prime factorization of any number. Within a factor tree, pairs of factors are found until no other factors can be used, as in the following factor tree of the number 84:

A factor tree

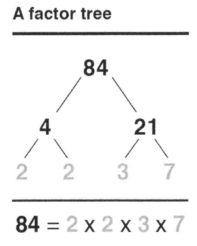

$$84 = 2 \times 2 \times 3 \times 7$$

It first breaks 84 into 21×4, which is not a prime factorization. Then, both 21 and 4 are factored into their primes. The final numbers on each branch consist of the numbers within the prime factorization. Therefore:

$$84 = 2 \times 2 \times 3 \times 7$$

Factorization can be helpful in finding greatest common divisors and least common denominators.

Also, a factorization of an algebraic expression can be found. Throughout the process, a more complicated expression can be decomposed into products of simpler expressions. To factor a polynomial, first determine if there is a greatest common factor. If there is, factor it out. For example, $2x^2 + 8x$ has a greatest common factor of $2x$ and can be written as $2x(x + 4)$. Once the greatest common monomial factor is factored out, if applicable, count the number of terms in the polynomial. If there are two terms, is it a difference of squares, a sum of cubes, or a difference of cubes?

If so, the following rules can be used:

Difference of Squares:
$$a^2 - b^2 = (a + b)(a - b)$$

Sum of Cubes:
$$a^3 + b^3 = (a + b)(a^2 - ab + b^2)$$

Difference of Cubes:
$$a^3 - b^3 = (a - b)(a^2 + ab + b^2)$$

If there are three terms, and if the trinomial is a perfect square trinomial, it can be factored into the following:
$$a^2 + 2ab + b^2 = (a + b)^2$$
$$a^2 - 2ab + b^2 = (a - b)^2$$

If not, try factoring into a product of two binomials in the form of:

$$(x + p)(x + q)$$

For example, to factor $x^2 + 6x + 8$, determine what two numbers have a product of 8 and a sum of 6. Those numbers are 4 and 2, so the trinomial factors into $(x + 2)(x + 4)$.

Finally, if there are four terms, try factoring by grouping. First, group terms together that have a common monomial factor. Then, factor out the common monomial factor from the first two terms. Next, look to see if a common factor can be factored out of the second set of two terms that results in a common binomial factor. Finally, factor out the common binomial factor of each expression, for example:

$$xy - x + 5y - 5 = x(y - 1) + 5(y - 1) = (y - 1)(x + 5)$$

After the expression is completely factored, check the factorization by multiplying it out; if this results in the original expression, then the factoring is correct. Factorizations are helpful in solving equations that consist of a polynomial set equal to 0. If the product of two algebraic expressions equals 0, then at least one of the factors is equal to 0. Therefore, factor the polynomial within the equation, set each factor equal to 0, and solve. For example, $x^2 + 7x - 18 = 0$ can be solved by factoring into:

$$(x + 9)(x - 2) = 0$$

Set each factor equal to 0, and solve to obtain $x = -9$ and $x = 2$.

Converting Non-Negative Fractions, Decimals, and Percentages

Within the number system, different forms of numbers can be used. It is important to be able to recognize each type, as well as work with, and convert between, the given forms. The **real number system** comprises natural numbers, whole numbers, integers, rational numbers, and irrational numbers. Natural numbers, whole numbers, integers, and irrational numbers typically are not represented as fractions, decimals, or percentages. Rational numbers, however, can be represented as any of these three forms.

A **rational number** is a number that can be written in the form $\frac{a}{b}$, where a and b are integers, and b is not equal to zero. In other words, rational numbers can be written in a fraction form. The value a is the **numerator**, and b is the **denominator**. If the numerator is equal to zero, the entire fraction is equal to zero. Non-negative fractions can be less than 1, equal to 1, or greater than 1. Fractions are less than 1 if the numerator is smaller (less than) than the denominator. For example, $\frac{3}{4}$ is less than 1. A fraction is equal to 1 if the numerator is equal to the denominator. For instance, $\frac{4}{4}$ is equal to 1. Finally, a fraction is greater than 1 if the numerator is greater than the denominator: the fraction $\frac{11}{4}$ is greater than 1.

When the numerator is greater than the denominator, the fraction is called an **improper fraction**. An improper fraction can be converted to a **mixed number**, a combination of both a whole number and a fraction. To convert an improper fraction to a mixed number, divide the numerator by the denominator. Write down the whole number portion, and then write any remainder over the original denominator. For example, $\frac{11}{4}$ is equivalent to $2\frac{3}{4}$. Conversely, a mixed number can be converted to an improper fraction by multiplying the denominator by the whole number and adding that result to the numerator.

Fractions can be converted to decimals. With a calculator, a fraction is converted to a decimal by dividing the numerator by the denominator. For example:

$$\frac{2}{5} = 2 \div 5 = 0.4$$

Sometimes, rounding might be necessary. Consider:

$$\frac{2}{7} = 2 \div 7 = 0.28571429$$

This decimal could be rounded for ease of use, and if it needed to be rounded to the nearest thousandth, the result would be 0.286. If a calculator is not available, a fraction can be converted to a decimal manually. First, find a number that, when multiplied by the denominator, has a value equal to 10, 100, 1,000, etc. Then, multiply both the numerator and denominator times that number. The decimal form of the fraction is equal to the new numerator with a decimal point placed as many place values to the left as there are zeros in the denominator.

For example, to convert $\frac{3}{5}$ to a decimal, multiply both the numerator and denominator times 2, which results in $\frac{6}{10}$. The decimal is equal to 0.6 because there is one zero in the denominator, and so the decimal place in the numerator is moved one unit to the left.

In the case where rounding would be necessary while working without a calculator, an approximation must be found. A number close to 10, 100, 1,000, etc. can be used. For example, to convert $\frac{1}{3}$ to a decimal, the numerator and denominator can be multiplied by 33 to turn the denominator into approximately 100,

55

which makes for an easier conversion to the equivalent decimal. This process results in $\frac{33}{99}$ and an approximate decimal of 0.33.

Once in decimal form, the number can be converted to a percentage. Multiply the decimal by 100 and then place a percent sign after the number. For example, 0.614 is equal to 61.4%. In other words, move the decimal place two units to the right and add the percent symbol.

Identifying Integers

Integers include zero, and both positive and negative numbers with no fractional component. Examples of integers are -3, 5, 120, -47, and 0. Numbers that are not integers include 1.3333, $\frac{1}{2}$, -5.7, and $4\frac{1}{2}$. Integers can be used to describe different real-world situations. If a scuba diver were to dive 50 feet down into the ocean, his position can be described as -50, in relation to sea level. If while traveling in Denver, Colorado, a car has an elevation reading of 2,300 feet, the integer 2,300 can be used to describe the feet above sea level. Integers can be used in many different ways to describe situations with whole numbers and zero.

Integers can also be added and subtracted as situations change. If the temperature in the morning is 45 degrees, and it dropped to 33 degrees in the afternoon, the difference can be found by subtracting the integers 45 and 33 to get a change of 12 degrees. If a submarine was at a depth of 100 feet below sea level, then rose 35 feet, the new depth can be found by adding -100 to 35. The following equation can be used to model the situation with integers:

$$-100 + 35 - 65$$

The answer of -65 reveals the new depth of the submarine, 65 feet below sea level.

Arithmetic Operations with Rational Numbers

The four basic operations include addition, subtraction, multiplication, and division. The result of addition is a sum, the result of subtraction is a difference, the result of multiplication is a product, and the result of division is a quotient. Each type of operation can be used when working with rational numbers; however, the basic operations need to be understood first while using simpler numbers before working with fractions and decimals.

These operations should first be learned using whole numbers. Addition needs to be done column by column. To add two whole numbers, add the ones column first, then the tens columns, then the hundreds, etc. If the sum of any column is greater than 9, a one must be carried over to the next column. For example, the following is the result of $482 + 924$:

$$
\begin{array}{r}
{}^{1} \\
482 \\
+924 \\
\hline
1406
\end{array}
$$

Notice that the sum of the tens column was 10, so a one was carried over to the hundreds column. Subtraction is also performed column by column. Subtraction is performed in the ones column first, then the tens, etc. If the number on top is less than the number below, a one must be borrowed from the column to the left. For example, the following is the result of $5,424 - 756$:

$$
\begin{array}{r}
4\ 13\ 11\ 14 \\
\cancel{5\ 4\ 2\ 4} \\
-\ 7\ 5\ 6 \\
\hline
4\ 6\ 6\ 8
\end{array}
$$

Notice that a one is borrowed from the tens, hundreds, and thousands place. After subtraction, the answer can be checked through addition. A check of this problem would be to show that $756 + 4,668 = 5,424$.

In multiplication, the number on top is known as the multiplicand, and the number below is the multiplier. Complete the problem by multiplying the multiplicand by each digit of the multiplier. Make sure to place the ones value of each result under the multiplying digit in the multiplier. The final product is found by adding each partial product. The following example shows the process of multiplying 46 times 37:

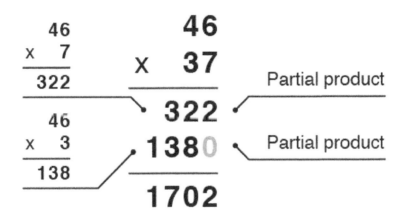

Finally, division can be performed using long division. When dividing, the first number is known as the **dividend**, and the second is the **divisor**. For example, with $a \div b = c$, a is the dividend, b is the divisor, and c is the quotient. For long division, place the dividend within the division bar and the divisor on the outside. For example, with $8,764 \div 4$, refer to the first problem in the diagram below. The first digit, 8, is divisible by 4 two times. Therefore, 2 goes above the division bar over the 8. Then, multiply 4 times 2 to get 8, and that product goes below the 8. Subtract to get 0, and then carry down the second digit, 7. Continue the same steps. $7 \div 4 = 1\ R\ 3$, so 1 is written above the 7. Multiply 4 times 1 to get 4, and write it below the 7. Subtract to get 3, and carry the 6 down next to the 3. Continuing this process for the next two digits results in a 9 and a 1. The final subtraction results in a 0, which means that 8,764 is evenly divisible by 4 with no remaining numbers.

The second example shows that:

$$4,536 \div 216 = 21$$

The steps are a little different because 216 cannot be contained in 4 or 5, so the first step is placing a 2 above the 3 because there are two 216's in 453. Finally, the third example shows that:

$$546 \div 31 = 17 \text{ R } 19$$

The 19 is a remainder.

Notice that the final subtraction does not result in a 0, which means that 546 is not divisible by 31. The remainder can also be written as a fraction over the divisor to say that:

$$546 \div 31 = 17\frac{19}{31}$$

$$
\begin{array}{r}
2191 \\
4\,\overline{|8764} \\
8 \\
\overline{07} \\
4 \\
\overline{36} \\
36 \\
\overline{04} \\
4 \\
\overline{0}
\end{array}
\qquad
\begin{array}{r}
21 \\
216\,\overline{|4536} \\
432 \\
\overline{216} \\
216 \\
\overline{0}
\end{array}
\qquad
\begin{array}{r}
17 \text{ r } 19 \\
31\,\overline{|546} \\
31 \\
\overline{236} \\
217 \\
\overline{19}
\end{array}
$$

A remainder can have meaning in a division problem with real-world application. For example, consider the third example:

$$546 \div 31 = 17 \text{ R } 19$$

Let's say that we had $546 to spend on calculators that cost $31 each, and we wanted to know how many we could buy. The division problem would answer this question. The result states that 17 calculators could be purchased, with $19 left over. Notice that the remainder will never be greater than or equal to the divisor.

Once the operations are understood with whole numbers, they can be used with negative numbers. There are many rules surrounding operations with negative numbers. First, consider addition with integers. The sum of two numbers can first be shown using a number line. For example, to add $-5 + (-6)$, plot the point -5 on the number line. Adding a negative number is the same as subtracting, so move 6 units to the left. This process results in landing on -11 on the number line, which is the sum of -5 and -6. If adding a positive number, move to the right. While visualizing this process using a number line is useful for understanding, it is more efficient to learn the rules of operations.

When adding two numbers with the same sign, add the absolute values of both numbers, and use the common sign of both numbers as the sign of the sum. For example, to add $-5 + (-6)$, add their absolute values:

$$5 + 6 = 11$$

Then, introduce a negative number because both addends are negative. The result is -11.

To add two integers with unlike signs, subtract the lesser absolute value from the greater absolute value, and apply the sign of the number with the greater absolute value to the result. For example, the sum $-7 + 4$ can be computed by finding the difference $7 - 4 = 3$ and then applying a negative because the value with the larger absolute value is negative. The result is -3. Similarly, the sum $-4 + 7$ can be found by computing the same difference but leaving it as a positive result because the addend with the larger absolute value is positive. Also, recall that any number plus 0 equals that number. This is known as the **Addition Property of 0**.

Subtracting two integers with opposite signs can be computed by changing to addition to avoid confusion. The rule is to add the first number to the opposite of the second number. The opposite of a number is the number with the same value on the other side of 0 on the number line. For example, -2 and 2 are opposites. Consider $4 - 8$. Change this to adding the opposite as follows: $4 + (-8)$. Then, follow the rules of addition of integers to obtain -4. Secondly, consider $-8 - (-2)$. Change this problem to adding the opposite as $-8 + 2$, which equals -6. Notice that subtracting a negative number functions the same as adding a positive number.

Multiplication and division of integers are actually less confusing than addition and subtraction because the rules are simpler to understand. If two factors in a multiplication problem have the same sign, the result is positive. If one factor is positive and one factor is negative, the result, known as the **product**, is negative. For example, $(-9)(-3) = 27$ and $9(-3) = -27$. Also, any number times 0 always results in 0.

If a problem consists of several multipliers, the result is negative if it contains an odd number of negative factors, and the result is positive if it contains an even number of negative factors. For example:

$$(-1)(-1)(-1)(-1) = 1 \text{ and } (-1)(-1)(-1)(-1)(-1) = -1$$

These two problems are also examples of repeated multiplication, which can be written in a more compact notation using exponents. The first example can be written as $(-1)^4 = 1$, and the second example can be written as $(-1)^5 = -1$. Both are exponential expressions; -1 is the base in both instances, and 4 and 5 are the respective exponents. Note that a negative number raised to an odd power is always negative, and a negative number raised to an even power is always positive. Also, $(-1)^4$ is not the same as -1^4. In the first expression, the negative is included in the parentheses, but it is not in the second expression. The second expression is found by evaluating 1^4 first to get 1 and then by applying the negative sign to obtain -1.

Similar rules apply within division. First, consider some vocabulary. When dividing 14 by 2, it can be written in the following ways: $14 \div 2 = 7$ or $\frac{14}{2} = 7$. 14 is the **dividend**, 2 is the **divisor**, and 7 is the **quotient**. If two numbers in a division problem have the same sign, the quotient is positive. If two numbers in a division problem have different signs, the quotient is negative. For example,

$$14 \div (-2) = -7, \text{ and } -14 \div (-2) = 7$$

To check division, multiply the quotient by the divisor to obtain the dividend. Also, remember that 0 divided by any number is equal to 0. However, any number divided by 0 is undefined. It just does not make sense to divide a number by 0 parts.

If more than one operation is to be completed in a problem, follow the **Order of Operations**. The mnemonic device, PEMDAS, states the order in which addition, subtraction, multiplication, and division need to be done. It also includes when to evaluate operations within grouping symbols and when to incorporate exponents. PEMDAS, which some remember by thinking "please excuse my dear Aunt Sally," refers to parentheses, exponents, multiplication, division, addition, and subtraction.

First, complete any operation within parentheses or any other grouping symbol like brackets, braces, or absolute value symbols. Note that this does not refer to when parentheses are used to represent multiplication like $(2)(5)$. In this case, an operation is not within parentheses like it is in (2×5). Then, any exponents must be computed. Next, multiplication and division are performed from left to right. Finally, addition and subtraction are performed from left to right. The following is an example in which the operations within the parentheses need to be performed first, so the order of operations must be applied to the exponent, subtraction, addition, and multiplication within the grouping symbol:

$$9 - 3(3^2 - 3 + 4 \cdot 3)$$

$$9 - 3(3^2 - 3 + 4 \cdot 3) \quad \text{Work within the parentheses first}$$

$$= 9 - 3(9 - 3 + 12)$$

$$= 9 - 3(18)$$

$$= 9 - 54$$

$$= -45$$

Once the rules for integers are understood, move on to learning how to perform operations with fractions and decimals. Recall that a rational number can be written as a fraction and can be converted to a decimal through division. If a rational number is negative, the rules for adding, subtracting, multiplying, and dividing integers must be used. If a rational number is in fraction form, performing addition, subtraction, multiplication, and division is more complicated than when working with integers. First, consider addition. To add two fractions having the same denominator, add the numerators and then reduce the fraction. When an answer is a fraction, it should always be in lowest terms. **Lowest terms** means that every common factor, other than 1, between the numerator and denominator is divided out.

For example:

$$\frac{2}{8} + \frac{4}{8} = \frac{6}{8} = \frac{6 \div 2}{8 \div 2} = \frac{3}{4}$$

Both the numerator and denominator of $\frac{6}{8}$ have a common factor of 2, so 2 is divided out of each number to put the fraction in lowest terms. If denominators are different in an addition problem, the fractions must be converted to have common denominators. The **least common denominator (LCD)** of all the given denominators must be found, and this value is equal to the **least common multiple (LCM)** of the denominators. This non-zero value is the smallest number that is a multiple of both denominators. Then, rewrite each original fraction as an equivalent fraction using the new denominator.

Once in this form, apply the process of adding with like denominators. For example, consider $\frac{1}{3} + \frac{4}{9}$. The LCD is 9 because it is the smallest multiple of both 3 and 9. The fraction $\frac{1}{3}$ must be rewritten with 9 as its denominator. Therefore, multiply both the numerator and denominator by 3. Multiplying by $\frac{3}{3}$ is the same as multiplying by 1, which does not change the value of the fraction. Therefore, an equivalent fraction is $\frac{3}{9}$, and $\frac{1}{3} + \frac{4}{9} = \frac{3}{9} + \frac{4}{9} = \frac{7}{9}$, which is in lowest terms. Subtraction is performed in a similar manner; once the denominators are equal, the numerators are then subtracted. The following is an example of addition of a positive and a negative fraction:

$$-\frac{5}{12} + \frac{5}{9} = -\frac{5 \times 3}{12 \times 3} + \frac{5 \times 4}{9 \times 4} = -\frac{15}{36} + \frac{20}{36} = \frac{5}{36}$$

Common denominators are not used in multiplication and division. To multiply two fractions, multiply the numerators together and the denominators together. Then, write the result in lowest terms. For example,

$$\frac{2}{3} \times \frac{9}{4} = \frac{18}{12} = \frac{3}{2}$$

Alternatively, the fractions could be factored first to cancel out any common factors before performing the multiplication. For example,

$$\frac{2}{3} \times \frac{9}{4} = \frac{2}{3} \times \frac{3 \times 3}{2 \times 2} = \frac{3}{2}$$

This second approach is helpful when working with larger numbers, as common factors might not be obvious. Multiplication and division of fractions are related because the division of two fractions is changed into a multiplication problem. This means that dividing a fraction by another fraction is the same as multiplying the first fraction by the reciprocal of the second fraction, so that second fraction must be inverted, or "flipped," to be in reciprocal form. For example,

$$\frac{11}{15} \div \frac{3}{5} = \frac{11}{15} \times \frac{5}{3} = \frac{55}{45} = \frac{11}{9}$$

The fraction $\frac{5}{3}$ is the reciprocal of $\frac{3}{5}$. It is possible to multiply and divide numbers containing a mix of integers and fractions. In this case, convert the integer to a fraction by placing it over a denominator of 1. For example, a division problem involving an integer and a fraction is:

$$3 \div \frac{1}{2} = \frac{3}{1} \times \frac{2}{1} = \frac{6}{1} = 6$$

Finally, when performing operations with rational numbers that are negative, the same rules apply as when performing operations with integers. For example, a negative fraction times a negative fraction results in a positive value, and a negative fraction subtracted from a negative fraction results in a negative value.

Operations can be performed on rational numbers in decimal form. Recall that to write a fraction as an equivalent decimal expression, divide the numerator by the denominator. For example,

$$\frac{1}{8} = 1 \div 8 = 0.125$$

With the case of decimals, it is important to keep track of place value. To add decimals, make sure the decimal places are in alignment and add vertically. If the numbers do not line up because there are extra or missing place values in one of the numbers, then zeros may be used as placeholders. For example, $0.123 + 0.23$ becomes:

$$\begin{array}{r} 0.123 \\ + \underline{0.230} \\ 0.353 \end{array}$$

Subtraction is done the same way. Multiplication and division are more complicated. To multiply two decimals, place one on top of the other as in a regular multiplication process and do not worry about lining up the decimal points. Then, multiply as with whole numbers, ignoring the decimals. Finally, in the solution, insert the decimal point as many places to the left as there are total decimal values in the original problem. Here is an example of a decimal multiplication problem:

$$\begin{array}{rl} 0.52 & \textit{2 decimal places} \\ \times \ \underline{0.2} & \textit{1 decimal place} \\ 0.104 & \textit{3 decimal places} \end{array}$$

The answer to 52 times 2 is 104, and because there are three decimal values in the problem, the decimal point is positioned three units to the left in the answer.

The decimal point plays an integral role throughout the whole problem when dividing with decimals. First, set up the problem in a long division format. If the divisor is not an integer, move the decimal to the right as many units as needed to make it an integer. The decimal in the dividend must be moved to the right the same number of places to maintain equality. Then, complete division normally. Here is an example of long division with decimals using the problem $12.72 \div 0.06$.

First, move the decimal point in 12.72 and 0.06 two places to the right. This gives 1,272 and 6.

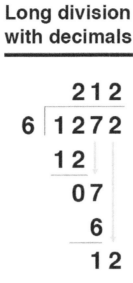

The decimal point in 0.06 needed to move two units to the right to turn it into an integer (6), so it also needed to move two units to the right in 12.72 to make it 1,272. The result is 212. Now move the decimal place back to the left by two units to get 2.12. This is the answer to $12.72 \div 0.06$. Also remember that a division problem can always be checked by multiplying the answer times the divisor to see if the result is equal to the dividend.

Sometimes it is helpful to round answers that are in decimal form. First, find the place to which the rounding needs to be done. Then, look at the digit to the right of it. If that digit is 4 or less, the number in the place value to its left stays the same, and everything to its right becomes a 0. This process is known as **rounding down**. If that digit is 5 or higher, the number in the place value to its left increases by 1, and every number to its right becomes a 0. This is called rounding up. Excess 0s at the end of a decimal can be dropped. For example, 0.145 rounded to the nearest hundredth place would be rounded up to 0.15, and 0.145 rounded to the nearest tenth place would be rounded down to 0.1.

Another operation that can be performed on rational numbers is the square root. Dealing with real numbers only, the positive square root of a number is equal to one of the two repeated positive factors of that number. For example:

$$\sqrt{49} = \sqrt{7 \times 7} = 7$$

A **perfect square** is a number that has a whole number as its square root. Examples of perfect squares are 1, 4, 9, 16, 25, etc. If a number is not a perfect square, an approximation can be used with a calculator. For example, $\sqrt{67} = 8.185$, rounded to the nearest thousandth place. Taking the square root of a fraction that includes perfect squares involves breaking up the problem into the square root of the numerator separate from the square root of the denominator.

For example,

$$\sqrt{\frac{16}{25}} = \frac{\sqrt{16}}{\sqrt{25}} = \frac{4}{5}$$

If the fraction does not contain perfect squares, a calculator can be used. Therefore, $\sqrt{\frac{2}{5}} = 0.632$, rounded to the nearest thousandth place. A common application of square roots involves the Pythagorean Theorem. Given a right triangle, the sum of the squares of the two legs equals the square of the hypotenuse.

For example, consider the following right triangle:

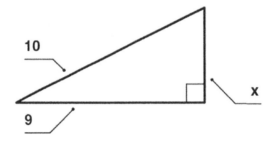

The missing side, x, can be found using the Pythagorean Theorem:

$$9^2 + x^2 = 10^2$$

$$81 + x^2 = 100$$

$$x^2 = 19$$

To solve for x, take the square root of both sides. Therefore, $x = \sqrt{19} = 4.36$, which has been rounded to two decimal places.

In addition to the square root, the cube root is another operation. If a number is a **perfect cube**, the cube root of that number is equal to one of the three repeated factors. For example:

$$\sqrt[3]{27} = \sqrt[3]{3 \times 3 \times 3} = 3$$

A negative number has a cube root, which will also be a negative number. For example:

$$\sqrt[3]{-27} = \sqrt[3]{(-3)(-3)(-3)} = -3$$

Similar to square roots, if the number is not a perfect cube, a calculator can be used to find an approximation. Therefore, $\sqrt[3]{\frac{2}{3}} = 0.873$, rounded to the nearest thousandth place.

Higher-order roots also exist. The number relating to the root is known as the **index**. Given the following root, $\sqrt[3]{64}$, 3 is the index, and 64 is the **radicand**. The entire expression is known as the **radical**. Higher-order roots exist when the index is larger than 3. They can be broken up into two groups: even and odd roots. Even roots, when the index is an even number, follow the properties of square roots. They are found by finding the number that, when multiplied by itself the number of times indicated by the index, results in the radicand. For example, the fifth root of 32 is equal to 2 because:

$$\sqrt[5]{32} = \sqrt[5]{2 \times 2 \times 2 \times 2 \times 2} = 2$$

Odd roots, when the index is an odd number, follow the properties of cube roots. A negative number has an odd root. Similarly, an odd root is found by finding the single factor that is repeated that many times to obtain the radicand. For example, the 4th root of 81 is equal to 3 because $3^4 = 81$. This radical is written as $\sqrt[4]{81} = 3$.

When performing operations in rational numbers, sometimes it might be helpful to round the numbers in the original problem to get a rough estimate of what the answer should be. For example, if you walked into a grocery store and had a $20 bill, you could round each item to the nearest dollar and add up all the items to make sure that you will have enough money when you check out. This process involves obtaining an estimation of what the exact total would be. In other situations, it might be helpful to round to the nearest $10 amount or $100 amount.

Front-end rounding might be helpful as well in many situations. In this type of rounding, the first digit of a number is rounded to the highest possible place value. Then, all digits following the first become 0. Consider a situation in which you are at the furniture store and want to estimate your total on three pieces of furniture that cost $434.99, $678.99, and $129.99. Front-end rounding would round these three amounts to $500, $700, and $200. Therefore, the estimate of your total would be $500 + $700 + $200 = $1,400, compared to the exact total of $1,243.97. In this situation, the estimate is not that far off the exact answer.

Rounding is useful both for approximation when an exact answer is not needed and for comparison when an exact answer is needed. For instance, if you had a complicated set of operations to complete and your estimate was $1,000, but you obtained an exact answer of $100,000, then you know something is off. You might want to check your work to see if you made a mistake because an estimate should not be that different from an exact answer. Estimates can also be helpful with square roots. If the square root of a number is unknown, then you can use the closest perfect square to help you approximate. For example, $\sqrt{50}$ is not equal to a whole number, but 50 is close to 49, which is a perfect square, and $\sqrt{49} = 7$. Therefore, $\sqrt{50}$ is a little bit larger than 7. The actual approximation, rounded to the nearest thousandth, is 7.071.

Ordering and Comparing Rational Numbers

Ordering rational numbers is a way to compare two or more different numerical values. Determining whether two amounts are equal, less than, or greater than is the basis for comparing both positive and negative numbers. Also, a group of numbers can be compared by ordering them from the smallest amount to the largest amount. A few symbols are necessary to use when ordering rational numbers. The equals sign, =, shows that the two quantities on either side of the symbol have the same value. For example,

$$\frac{12}{3} = 4$$

because both values are equivalent. Another symbol that is used to compare numbers is <, which represents "less than." With this symbol, the smaller number is placed on the left and the larger number is placed on the right. Always remember that the symbol's "mouth" opens up to the larger number. When comparing negative and positive numbers, it is important to remember that the number occurring to the left on the number line is always smaller and is placed to the left of the symbol. This idea might seem confusing because some values could appear at first glance to be larger, even though they are not. For example, −5 < 4 is read "negative 5 is less than 4." Here is an image of a number line for help:

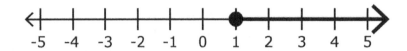

The symbol ≤ represents "less than or equal to," and it joins < with equality. Therefore, both −5 ≤ 4 and −5 ≤ −5 are true statements and "−5 is less than or equal to both 4 and −5." Other symbols are > and ≥, which represent "greater than" and "greater than or equal to." Both 4 ≥ −1 and −1 ≥ −1 are correct ways to use these symbols.

Here is a chart of these four inequality symbols:

Symbol	Definition
<	less than
≤	less than or equal to
>	greater than
≥	greater than or equal to

Comparing integers is a straightforward process, especially when using the number line, but the comparison of decimals and fractions is not as obvious. When comparing two non-negative decimals, compare digit by digit, starting from the left. The larger value contains the first larger digit. For example, 0.1456 is larger than 0.1234 because the value 4 in the hundredths place in the first decimal is larger than the value 2 in the hundredths place in the second decimal. When comparing a fraction with a decimal, convert the fraction to a decimal and then compare in the same manner.

Finally, there are a few options when comparing fractions. If two non-negative fractions have the same denominator, the fraction with the larger numerator is the larger value. If they have different denominators, they can be converted to equivalent fractions with a common denominator to be compared, or they can be converted to decimals to be compared.

When comparing two negative decimals or fractions, a different approach must be used. It is important to remember that the smaller number exists to the left on the number line. Therefore, when comparing two negative decimals by place value, the number with the larger first place value is smaller due to the negative sign. Whichever value is closer to 0 is larger. For instance, −0.456 is larger than −0.498 because of the values in the hundredth places. If two negative fractions have the same denominator, the fraction with the larger numerator is smaller because of the negative sign.

Applying Estimation Strategies and Rounding Rules to Real-World Problems

Sometimes it is helpful to find an estimated answer to a problem rather than working out an exact answer. An estimation might be much quicker to find, and it might be all that is required given the scenario. For example, if Aria goes grocery shopping and has only a $100 bill to cover all her purchases, it might be appropriate for her to estimate the total of the items she is purchasing to determine if she has enough money to cover them. Also, an estimation can help determine if an answer makes sense. For instance, if you estimate that an answer should be in the 100s, but your result is a fraction less than 1, something is probably wrong in the calculation.

The first type of estimation involves rounding. As mentioned, **rounding** consists of expressing a number in terms of the nearest decimal place like the tenth, hundredth, or thousandth place, or in terms of the nearest whole number unit like tens, hundreds, or thousands place. When rounding to a specific place value, look at the digit to the right of the place. If it is 5 or higher, round the number to its left up to the next value, and if it is 4 or lower, keep that number at the same value. For instance, 1,654.2674 rounded to the nearest thousand is 2,000, and the same number rounded to the nearest thousandth is 1,654.267. Rounding can make it easier to estimate totals at the store. Items can be rounded to the nearest dollar. For example, a can of corn that costs $0.79 can be rounded to $1.00, and then all other items can be rounded in a similar manner and added together.

When working with larger numbers, it might make more sense to round to higher place values. For example, when estimating the total value of a dealership's car inventory, it would make sense to round the car values to the nearest thousands place. The price of a car that is on sale for $15,654 can be estimated at $16,000. All other cars on the lot could be rounded in the same manner and then added together.

Depending on the situation, it might make sense to calculate an over-estimate. For example, to make sure Aria has enough money at the grocery store, rounding up for each item would ensure that she will have enough money when it comes time to pay. A $0.40 item rounded up to $1.00 would ensure that there is a dollar to cover that item. Traditional rounding rules would round $0.40 to $0, which does not make sense in this particular real-world setting. Aria might not have a dollar available at checkout to pay for that item if she uses traditional rounding. It is up to the customer to decide the best approach when estimating.

Estimating is also very helpful when working with measurements. Bryan is updating his kitchen and wants to retile the floor. Again, an over-measurement might be useful. Also, rounding to nearest half-unit might be helpful. For instance, one side of the kitchen might have an exact measurement of 14.32 feet, and the most useful measurement needed to buy tile could be estimating this quantity to be 14.5 feet. If the kitchen was rectangular and the other side measured 10.9 feet, Bryan might round the other side to 11 feet. Therefore, Bryan would find the total tile necessary according to the following area calculation: $14.5 \times 11 = 159.5$ square feet. To make sure he purchases enough tile, Bryan would probably want to purchase at least 160 square feet of tile. This is a scenario in which an estimation might be more useful than an exact calculation. Having more tile than necessary is better than having an exact amount, in case any tiles are broken or otherwise unusable.

Finally, estimation is helpful when exact answers are necessary. Consider a situation in which Sabina has many operations to perform on numbers with decimals, and she is allowed a calculator to find the result. Even though an exact result can be obtained with a calculator, there is always a possibility that Sabina could make an error while inputting the data. For example, she could miss a decimal place, or misuse a parenthesis, causing a problem with the actual order of operations. A quick estimation at the beginning could help ensure that her final answer is within the correct range.

Sabina has to find the exact total of 10 cars listed for sale at the dealership. Each price has two decimal places included to account for both dollars and cents. If one car is listed at $21,234.43 but Sabina incorrectly inputs into the calculator the price of $2,123.443, this error would throw off the final sum by almost $20,000. A quick estimation at the beginning, by rounding each price to the nearest thousands place and finding the sum of the prices, would give Sabina an amount to compare the exact amount to. This comparison would let Sabina see if an error was made in her exact calculation.

Percentages

Percentages are defined as parts per one hundred. To convert a decimal to a percentage, move the decimal point two units to the right and place the percent sign after the number. Percentages appear in many scenarios in the real world. It is important to make sure the statement containing the percentage is translated to a correct mathematical expression. Be aware that it is extremely common to make a mistake when working with percentages within word problems.

An example of a word problem containing a percentage is the following: 35% of people speed when driving to work. In a group of 5,600 commuters, how many would be expected to speed on the way to their place of employment? The answer to this problem is found by finding 35% of 5,600. First, change the percentage to the decimal 0.35. Then compute the product:

$$0.35 \times 5,600 = 1,960$$

Therefore, it would be expected that 1,960 of those commuters would speed on their way to work based on the data given. In this situation, the word "of" signals to use multiplication to find the answer.

Another way percentages are used is in the following problem: Teachers work 8 months out of the year. What percent of the year do they work? To answer this problem, find what percent of 12 the number 8 is, because there are 12 months in a year. Therefore, divide 8 by 12, and convert that number to a percentage:

$$\frac{8}{12} = \frac{2}{3} = 0.66\overline{6}$$

The percentage rounded to the nearest tenth place tells us that teachers work 66.7% of the year. Percentage problems can also find missing quantities like in the following question: 60% of what number is 75? To find the missing quantity, turn the question into an equation. Let x be equal to the missing quantity. Therefore, $0.60x = 75$. Divide each side by 0.60 to obtain 125. Therefore, 60% of 125 is equal to 75.

Sales tax is an important application relating to percentages because tax rates are usually given as percentages. For example, a city might have an 8% sales tax rate. Therefore, when an item is purchased with that tax rate, the real cost to the customer is 1.08 times the price in the store. For example, a $25 pair of jeans costs the customer:

$$\$25 \times 1.08 = \$27$$

If the sales tax rate is unknown, it can be determined after an item is purchased. If a customer visits a store and purchases an item for $21.44, but the price in the store was $19, they can find the tax rate by first subtracting:

$$\$21.44 - \$19$$

to obtain $2.44, the sales tax amount. The sales tax is a percentage of the in-store price. Therefore, the tax rate is:

$$\frac{2.44}{19} = 0.128$$

which has been rounded to the nearest thousandths place. In this scenario, the actual sales tax rate given as a percentage is 12.8%.

68

Proportions

Fractions appear in everyday situations, and in many scenarios, they appear in the real-world as ratios and in proportions. A **ratio** is formed when two different quantities are compared. For example, in a group of 50 people, if there are 33 females and 17 males, the ratio of females to males is 33 to 17. This expression can be written in the fraction form as $\frac{33}{50}$, where the denominator is the sum of females and males, or by using the ratio symbol, $33 : 17$. The order of the number matters when forming ratios. In the same setting, the ratio of males to females is 17 to 33, which is equivalent to $\frac{17}{50}$ or $17 : 33$.

A **proportion** is an equation involving two ratios. The equation $\frac{a}{b} = \frac{c}{d}$ or $a : b = c : d$ is a proportion, for real numbers a, b, c, and d. Usually, in one ratio, one of the quantities is unknown, and cross-multiplication is used to solve for the unknown. Consider:

$$\frac{1}{4} = \frac{x}{5}$$

To solve for x, cross-multiply to obtain $5 = 4x$. Divide each side by 4 to obtain the solution:

$$x = \frac{5}{4}$$

It is also true that percentages are ratios in which the second term is 100 minus the first term. For example, 65% is 65:35 or $\frac{65}{100}$. Therefore, when working with percentages, one is also working with ratios.

Real-world problems frequently involve proportions. For example, consider the following problem: If 2 out of 50 pizzas are usually delivered late from a local Italian restaurant, how many would be late out of 235 pizzas? The following proportion would be solved with x as the unknown quantity of late pizza:

$$\frac{2}{50} = \frac{x}{235}$$

Cross-multiplying results in $470 = 50x$. Divide both sides by 50 to obtain $x = \frac{470}{50}$, which in lowest terms is equal to $\frac{47}{5}$. In decimal form, this improper fraction is equal to 9.4. Because it does not make sense to answer this question with decimals (portions of pizza do not get delivered) the answer must be rounded. Traditional rounding rules would say that 9 pizzas would be expected to be delivered late. However, to be safe, rounding up to 10 pizzas out of 235 would probably make more sense.

Ratios and Rates of Change

Recall that a **ratio** is the comparison of two different quantities. Comparing 2 apples to 3 oranges results in the ratio $2 : 3$, which can be expressed as the fraction $\frac{2}{5}$. Note that order is important when discussing ratios. The number mentioned first is the antecedent, and the number mentioned second is the consequent. Note that the consequent of the ratio and the denominator of the fraction are *not* the same. When there are 2 apples to 3 oranges, there are five fruit total; two fifths of the fruit are apples, while three fifths are oranges. The ratio $2 : 3$ represents a different relationship than the ratio $3 : 2$.

Also, it is important to make sure that when discussing ratios that have units attached to them, the two quantities use the same units. For example, to think of 8 feet to 4 yards, it would make sense to convert 4 yards to feet by multiplying by 3. Therefore, the ratio would be 8 feet to 12 feet, which can be expressed as the fraction $\frac{8}{20}$. Also, note that it is proper to refer to ratios in lowest terms. Therefore, the ratio of 8 feet to 4 yards is equivalent to the fraction $\frac{2}{5}$.

Many real-world problems involve ratios. Often, problems with ratios involve proportions, as when two ratios are set equal to find the missing amount. However, some problems involve deciphering single ratios. For example, consider an amusement park that sold 345 tickets last Saturday. If 145 tickets were sold to adults and the rest of the tickets were sold to children, what would the ratio of the number of adult tickets to children's tickets be? A common mistake would be to say the ratio is $145 : 345$. However, 345 is the total number of tickets sold, not the number of children's tickets. There were $345 - 145 = 200$ tickets sold to children. The correct ratio of adult to children's tickets is $145 : 200$. As a fraction, this expression is written as $\frac{145}{345}$, which can be reduced to $\frac{29}{69}$.

While a ratio compares two measurements using the same units, rates compare two measurements with different units. Examples of rates would be $200 for 8 hours of work, or 500 miles traveled per 20 gallons. Because the units are different, it is important to always include the units when discussing rates. Key words in rate problems include for, per, on, from, and in. Just as with ratios, it is important to write rates in lowest terms. A common rate in real-life situations is cost per unit, which describes how much one item/unit costs. When evaluating the cost of an item that comes in several sizes, the cost per unit rate can help buyers determine the best deal.

For example, if 2 quarts of soup was sold for $3.50 and 3 quarts was sold for $4.60, to determine the best buy, the cost per quart should be found. $\frac{\$3.50}{2 \text{ qt}} = \$1.75$ per quart, and:

$$\frac{\$4.60}{3 \text{ qt}} = \$1.53 \text{ per quart}$$

Therefore, the better deal would be the 3-quart option.

Rate of change problems involve calculating a quantity per some unit of measurement. Usually, the unit of measurement is time. For example, meters per second is a common rate of change. To calculate this measurement, find the amount traveled in meters and divide by total time traveled. The result is the average speed over the entire time interval. Another common rate of change used in the real world is miles per hour.

Consider the following problem that involves calculating an average rate of change in temperature. Last Saturday, the temperature at 1:00 a.m. was 34 degrees Fahrenheit, and at noon, the temperature had increased to 75 degrees Fahrenheit. What was the average rate of change over that time interval? The average rate of change is calculated by finding the change in temperature and dividing by the total hours elapsed. Therefore, the rate of change was equal to $\frac{75-34}{12-1} = \frac{41}{11}$ degrees per hour. This quantity rounded to two decimal places is equal to 3.73 degrees per hour.

A common rate of change that appears in algebra is the slope calculation. Given a linear equation in one variable, $y = mx + b$, the **slope**, m, is equal to:

$$\frac{rise}{run}$$

or

$$\frac{change\ in\ y}{change\ in\ x}$$

In other words, slope is equivalent to the ratio of the vertical and horizontal changes between any two points on a line. The vertical change is known as the **rise**, and the horizontal change is known as the **run**. Given any two points on a line (x_1, y_1) and (x_2, y_2), slope can be calculated with the formula:

$$m = \frac{y_2 - y_1}{x_2 - x_1} = \frac{\Delta y}{\Delta x}$$

Common real-world applications of slope include determining how steep a staircase should be, calculating how steep a road is, and determining how to build a wheelchair ramp.

Many times, problems involving rates and ratios involve proportions. A proportion states that two ratios (or rates) are equal. The property of cross products can be used to determine if a proportion is true, meaning both ratios are equivalent. If $\frac{a}{b} = \frac{c}{d}$, then to clear the fractions, multiply both sides by the least common denominator, bd. This results in $ad = bc$, which is equal to the result of multiplying along both diagonals. For example, $\frac{4}{40} = \frac{1}{10}$ grants the cross product $4 \times 10 = 40 \times 1$ which is equivalent to $40 = 40$ and shows that this proportion is true. Cross products are used when proportions are involved in real-world problems. Consider the following: If 3 pounds of fertilizer will cover 75 square feet of grass, how many pounds are needed for 375 square feet? To solve this problem, set up a proportion using two ratios. Let x equal the unknown quantity, pounds needed for 375 feet. Then, the equation found by setting the two given ratios equal to one another is $\frac{3}{75} = \frac{x}{375}$.

Cross multiplication gives:

$$3 \times 375 = 75x$$

Therefore, $1,125 = 75x$. Divide both sides by 75 to get $x = 15$. Therefore, 15 gallons of fertilizer is needed to cover 375 square feet of grass.

Another application of proportions involves similar triangles. If two triangles have corresponding angles with the same measurements and corresponding sides with proportional measurements, the triangles are said to be similar. If two angles are the same, the third pair of angles are equal as well because the sum of all angles in a triangle is equal to 180 degrees. Each pair of equivalent angles are known as **corresponding angles**. **Corresponding sides** face the corresponding angles, and it is true that corresponding sides are in proportion.

For example, consider the following set of similar triangles:

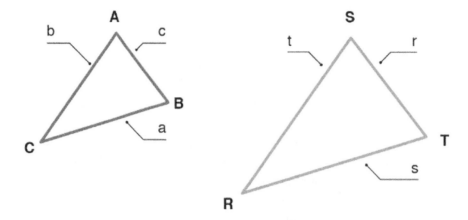

Angles A and S have the same measurement, angles C and R have the same measurement, and angles B and T have the same measurement. Therefore, the following proportion can be set up from the sides:

$$\frac{c}{r} = \frac{a}{s} = \frac{b}{t}$$

This proportion can be helpful in finding missing lengths in pairs of similar triangles. For example, if the following triangles are similar, a proportion can be used to find the missing side lengths, a and b.

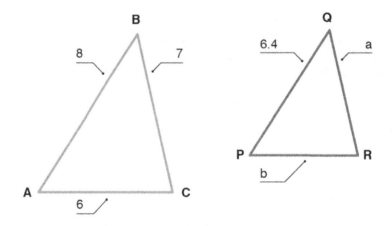

The proportions $\frac{8}{6.4} = \frac{6}{b}$ and $\frac{8}{6.4} = \frac{7}{a}$ can both be cross-multiplied and solved to obtain $a = 5.6$ and $b = 4.8$.

A real-life situation that uses similar triangles involves measuring shadows to find heights of unknown objects. Consider the following problem: A building casts a shadow that is 120 feet long, and at the same time, another building that is 80 feet high casts a shadow that is 60 feet long. How tall is the first building? Each building, together with the sun rays and shadows casted on the ground, forms a triangle. They are similar because each building forms a right angle with the ground, and the sun rays form equivalent angles. Therefore, these two pairs of angles are both equal. Because all angles in a triangle add up to 180 degrees, the third angles are equal as well. Both shadows form corresponding sides of the triangle, the buildings form corresponding sides, and the sun rays form corresponding sides. Therefore, the triangles are similar, and the following proportion can be used to find the missing building length:

$$\frac{120}{x} = \frac{60}{80}$$

Cross-multiply to obtain the equation $9,600 = 60x$. Then, divide both sides by 60 to obtain $x = 160$. This means that the first building is 160 feet high.

Solving Unit Rate Problems

A **unit rate** is a rate with a denominator of one. It is a comparison of two values with different units where one value is equal to one. Examples of unit rates include 60 miles per hour and 200 words per minute. Problems involving unit rates may require some work to find the unit rate. For example, if Mary travels 360 miles in 5 hours, what is her speed, expressed as a unit rate? The rate can be expressed as the following fraction: $\frac{360 \text{ miles}}{5 \text{ hours}}$. The denominator can be changed to one by dividing by five. The numerator will also need to be divided by five to follow the rules of equality. This division turns the fraction into $\frac{72 \text{ miles}}{1 \text{ hour}}$, which can now be labeled as a unit rate because one unit has a value of one.

Another type of question involves the use of unit rates to solve problems. For example, if Trey needs to read 300 pages and his average speed is 75 pages per hour, will he be able to finish the reading in 5 hours? The unit rate is 75 pages per hour, so the total of 300 pages can be divided by 75 to find the time. After the division, the time it takes to read is four hours. The answer to the question is yes, Trey will finish the reading within 5 hours.

Recognizing Rational Exponents

Rational exponents are used to express the root of a number raised to a specific power. For example, $3^{\frac{1}{2}}$ has a base of 3 and rational exponent of $\frac{1}{2}$. The square root of 3 raised to the first power can be written as $\sqrt[2]{3^1}$. Any number with a rational exponent can be written this way. The numerator, or number on top of the fraction, becomes the whole number exponent and the denominator, or bottom number of the fraction, becomes the root. Another example is $4^{\frac{3}{2}}$. It can be rewritten as the square root of four to the third power, or $\sqrt[2]{4^3}$. To simplify this, first solve for 4 to the third power:

$$4^3 = 4 \times 4 \times 4 = 64$$

Then take the square root of 64, written as $\sqrt[2]{64}$, which yields an answer of 8. Another way of stating the answer would be 4 to the power of $\frac{3}{2}$ is eight, or that 4 to the power of $\frac{3}{2}$ is the square root of 4 cubed:

$$\sqrt[2]{4}^3 = 2^3 = 2 \times 2 \times 2 = 8$$

Whole-Number Exponents

Numbers can also be written using exponents. The number 7,000 can be written as $7 \times 1,000$ because 7 is in the thousands place. It can also be written as 7×10^3 because $1,000 = 10^3$. Another number that can use this notation is 500. It can be written as 5×100, or 5×10^2, because $100 = 10^2$.

The number 30 can be written as 3×10, or 3×10^1, because $10 = 10^1$. Notice that each one of the exponents of 10 is equal to the number of zeros in the number. Seven is in the thousands place, with three zeros, and the exponent on ten is 3. The five is in the hundreds place, with two zeros, and the exponent on the ten is 2. A question may give the number 40,000 and ask for it to be rewritten using exponents with a base of ten. Because the number has a four in the ten-thousands place and four zeros, it can be written using an exponent of four: 4×10^4.

Understanding Vectors

A **vector** is something that has both magnitude and direction. A vector may sometimes be represented by a ray that has a length, for its magnitude, and a direction. As the magnitude of the vector increases, the length of the ray changes. The direction of the ray refers to the way that the magnitude is applied. The following vector shows the placement and parts of a vector:

Parts of a Vector

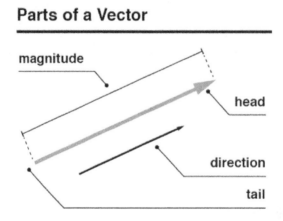

Examples of vectors include force and velocity. Force is a vector because applying force requires magnitude, which is the amount of force, and a direction, which is the way a force is applied. **Velocity** is a vector because it has a magnitude, or speed that an object travels, and also the direction that the object is going in. Vectors can be added together by placing the tail of the second at the head of the first. The resulting vector is found by starting at the first tail and ending at the second head.

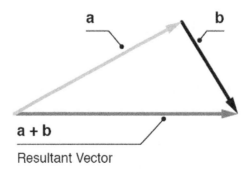

Resultant Vector

Subtraction can also be done with vectors by adding the inverse of the second vector. The inverse is found by reversing the direction of the vector.

Then addition can take place just as described earlier but using the inverse instead of the original vector. Scalar multiplication can also be done with vectors. This multiplication changes the magnitude of the vector by the scalar, or number. For example, if the length is described as 4, then scalar multiplication is used to multiply by 2, where the vector magnitude becomes 8. The direction of the vector is left unchanged because scalar does not include direction.

Vectors may also be described using coordinates on a plane, such as $(5, 2)$. This vector would start at a point and move to the right 5 and up 2. The two coordinates describe the horizontal and vertical components of the vector. The starting point in relation to the coordinates is the tail, and the ending point is the head.

Creating Matrices

A **matrix** is an arrangement of numbers in rows and columns. Matrices are used to work with vectors and transform them. One example is a system of linear equations. Matrices can represent a system and be used to transform and solve the system. An important connection between scalars, vectors, and matrices is this: scalars are only numbers, vectors are numbers with magnitude and direction, and matrices are an array of numbers in rows and columns. The rows run from left to right and the columns run from top to bottom. When describing the dimensions of a matrix, the number of rows is stated first, and the number of columns is stated second. The following matrix has two rows and three columns, referred to as a 2×3 matrix:

$$\begin{bmatrix} 3 & 5 & 7 \\ 4 & 2 & 8 \end{bmatrix}$$

A number in a matrix can be found by describing its location. For example, the number in row two, column three is 8. In row one, column two, the number 5 is found.

Operations can be performed on matrices, just as they can on vectors. Scalar multiplication can be performed on matrices and it will change the magnitude, just as with a vector. A scalar multiplication problem using a 2×2 matrix looks like the following:

$$3 \times \begin{bmatrix} 4 & 5 \\ 8 & 3 \end{bmatrix}$$

The scalar of 3 is multiplied by each number to form the resulting matrix: $\begin{bmatrix} 12 & 15 \\ 24 & 9 \end{bmatrix}$. Matrices can also be added and subtracted. For these operations to be performed, the matrices must be the same dimensions. Other operations that can be performed to manipulate matrices are multiplication, division, and transposition. **Transposing** a matrix means to switch the rows and columns. If the original matrix has two rows and three columns, then the transposed matrix has three rows and two columns.

Operations and Properties of Rational Numbers

Using Inverse Operations to Solve Problems

Inverse operations can be used to solve problems where there is a missing value. The area for a rectangle may be given, along with the length, but the width may be unknown. This situation can be modeled by the equation Area = Length × Width. The area is 40 square feet and the length is 10 feet. The equation becomes $40 = 10 \times w$. In order to find the w, we recognize that some number multiplied by 10 yields the number 40. The inverse operation to multiplication is division, so the 10 can be divided on both sides of the equation. This operation cancels out the 10 and yields an answer of 4 for the width. The following equation shows the work:

$$40 = 10 \times w$$

$$\frac{40}{10} = \frac{10 \times w}{10}$$

$$4 = w$$

Other inverse operations can be used to solve problems as well. The following equation can be solved for b:

$$b + 4 = 9$$

Because 4 is added to b, it can be subtracted on both sides of the equal sign to cancel out the four and solve for b, as follows:

$$b + 4 - 4 = 9 - 4$$

$$b = 5$$

Whatever operation is used in the equation, the inverse operation can be used and applied to both sides of the equals sign to solve for an unknown value.

Interpreting Remainders in Division Problems

Understanding remainders begins with understanding the division problem. The problem $24 \div 7$ can be read as "twenty-four divided by seven." The problem is asking how many groups of 7 will fit into 24. Counting by seven, the multiples are 7, 14, 21, 28. Twenty-one, which is three groups of 7, is the closest to 24. The difference between 21 and 24 is 3, which is called the remainder. This is a remainder because it is the number that is left out after the three groups of seven are taken from 24.

The answer to this division problem can be written as 3 with a remainder 3, or $3\frac{3}{7}$. The fraction $\frac{3}{7}$ can be used because it shows the part of the whole left when the division is complete. Another division problem may have the following numbers: $36 \div 5$. This problem is asking how many groups of 5 will fit evenly into 36. When counting by multiples of 5, the following list is generated: 5, 10, 15, 20, 25, 30, 35, 40. As seen in the list, there are seven groups of five that make 35. To get to the total of 36, there needs to be one additional number. The answer to the division problem would be $36 \div 5 = 7$ R 1, or $7\frac{1}{5}$. The fractional part represents the number that cannot make up a whole group of five.

Performing Operations on Rational Numbers

Rational numbers are any numbers that can be written as a fraction of integers. Operations to be performed on rational numbers include adding, subtracting, multiplying, and dividing. Essentially, this refers to performing these operations on fractions. For the denominators to become 35, the first fraction must be multiplied by 7 and the second by 5. For example, the problem $\frac{3}{5} + \frac{6}{7}$ requires that the common multiple be found between 5 and 7. The smallest number that divides evenly by 5 and 7 is 35. For the denominators to become 35, the first fraction must be multiplied by 7 and the second by 5. The fraction $\frac{3}{5}$ can be multiplied by 7 on the top and bottom to yield the fraction $\frac{21}{35}$. The fraction $\frac{6}{7}$ can be multiplied by 5 to yield the fraction $\frac{30}{35}$. Now that the fractions have the same denominator, the numerators can be added. The answer to the addition problem becomes:

$$\frac{3}{5} + \frac{6}{7} = \frac{21}{35} + \frac{30}{35} = \frac{51}{35}$$

The same technique can be used to subtract fractions. Multiplication and division may seem easier to perform because finding common denominators is unnecessary. If the problem reads $\frac{1}{3} \times \frac{4}{5}$, then the numerators and denominators are multiplied by each other and the answer is found to be $\frac{4}{15}$. For division, the problem must be changed to multiplication before performing operations. To complete a fraction division problem, you need to leave, change, and flip before multiplying. If the problem reads $\frac{3}{7} \div \frac{3}{4}$, then the first fraction is *left* alone, the operation is *changed* to multiplication, and then the last fraction is *flipped*.

The problem becomes:

$$\frac{3}{7} \times \frac{4}{3} = \frac{12}{21}$$

Rational numbers can also be negative. When two negative numbers are added, the result is a negative number with an even greater magnitude. When a negative number is added to a positive number, the result depends on the value of each addend. For example, $-4 + 8 = 4$ because the positive number is larger than the negative number. For multiplying two negative numbers, the result is positive. For example, $-4 \times -3 = 12$, where the negatives cancel out and yield a positive answer.

Rational Numbers and Their Operations

Rational numbers can be whole or negative numbers, fractions, or repeating decimals because these numbers can all be written as fractions. Whole numbers can be written as fractions; for example, 25 and 17 can be written as $\frac{25}{1}$ and $\frac{17}{1}$. One way of interpreting these fractions is to say that they are **ratios**, or comparisons of two quantities. The fractions given may represent 25 students to 1 classroom, or 17 desks to 1 computer lab. Repeating decimals can also be written as fractions of integers, such as 0.3333 and 0.6666667.

These repeating decimals can be written as the fractions $\frac{1}{3}$ and $\frac{2}{3}$. Fractions can be described as having a part-to-whole relationship. The fraction $\frac{1}{3}$ may represent 1 piece of pizza out of the whole cut into 3 pieces. The fraction $\frac{2}{3}$ may represent 2 pieces of the same whole pizza. One operation to perform on rational numbers, or fractions, can be addition. Adding the fractions $\frac{1}{3}$ and $\frac{2}{3}$ can be as simple as adding the numerators, 1 and 2. Because the denominator on both fractions is 3, that means the parts on the top have the same meaning or, in this case, are the same size piece of pizza. When adding these fractions, the result is $\frac{3}{3}$, or 1. Both numbers are rational and represent a whole, or in this problem, a whole pizza.

Other than fractions, rational numbers also include whole numbers and negative integers. When whole numbers are added, the result is always greater than the addends (unless 0 is added to the number, in which case its value would remain the same). For example, the equation $4 + 18 = 22$ shows 4 increased by 18, with a result of 22. When subtracting rational numbers, sometimes the result is a negative number. For example, the equation $5 - 12 = -7$ shows that taking 12 away from 5 results in a negative answer (–7) because the starting number (5) is smaller than the number taken away (12). For multiplication and division, similar results are found. Multiplying rational numbers may look like the following equation: $5 \times 7 = 35$, where both numbers are positive and whole, and the result is a larger number than the factors. The number 5 is counted 7 times, which results in a total of 35. Sometimes, the equation looks like $-4 \times 3 = -12$, so the result is negative because a positive number times a negative number gives a negative answer. The rule is that any time a negative number and a positive number are multiplied or divided, the result is negative.

Examples Where Multiplication Does Not Result in a Product Greater than Both Factors and Division Does Not Result in a Quotient Smaller than the Dividend

A common misconception of multiplication is that it always results in a value greater than the beginning number, or factors. This is not always the case. When working with fractions, multiplication may be used to take part of another number. For example, $\frac{1}{2} \times \frac{1}{4}$ can be read as "one-half times one-fourth," or taking one-half of one-fourth. The latter translation makes it easier to understand the concept. Taking half of one-fourth will result in a smaller number than one-fourth. It will result in one-eighth. The same happens with multiplying two-thirds times three-fifths, or $\frac{2}{3} \times \frac{3}{5}$. The concept of taking two-thirds, which is a part, of three-fifths, means that there will be an even smaller part as the result. Multiplication of these two fractions yields the answer $\frac{6}{15}$, or $\frac{2}{5}$.

In the same way, another misconception is that division always has results smaller than the beginning number or dividend. When working with whole numbers, division asks how many times a whole goes into another whole. This result will always be smaller than the dividend, where $6 \div 2 = 3$ and $20 \div 5 = 4$. When working with fractions, the number of times a part goes into another part depends on the value of each fraction. For example, three-fourths divided by one-fourth, or $\frac{3}{4} \div \frac{1}{4}$, asks to find how many times $\frac{1}{4}$ will go into $\frac{3}{4}$. Because these have the same denominator, the numerators can be compared as is, without needing to convert the fractions. The result is easily found to be 3 because one goes into three 3 times.

Composing and Decomposing Fractions

Fractions are ratios of whole numbers and their negatives. Fractions represent parts of wholes, whether pies, or money, or work. The number on top, or numerator, represents the part, and the bottom number, or denominator, represents the whole. The number $\frac{1}{2}$ represents half of a whole. Other ways to represent one-half are $\frac{2}{4}, \frac{3}{6}$, and $\frac{5}{10}$. These are fractions not written in simplest form, but the numerators are all halves of the denominators. The fraction $\frac{1}{4}$ represents 1 part to a whole of 4 parts.

This can be modeled by the quarter's value in relation to the dollar. One quarter is $\frac{1}{4}$ of a dollar. In the same way, 2 quarters make up $\frac{1}{2}$ of a dollar, so 2 fractions of $\frac{1}{4}$ make up a fraction of $\frac{1}{2}$. Three quarters make up three-fourths of a dollar. The three fractions of $\frac{1}{4} + \frac{1}{4} + \frac{1}{4}$ are equal to $\frac{3}{4}$ of a whole. This illustration can be seen using the bars below divided into one whole, then two halves, then three sections of one-third, then four sections of one-fourth. Based on the size of the fraction, different numbers of each fraction are needed to make up a whole.

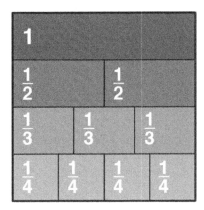

Value of a Unit Fraction

A **unit fraction** is a fraction where the numerator has a value of one. The fractions one-half, one-third, one-seventh, and one-tenth are all examples of unit fractions. Examples that are not unit fractions include three-fourths, four-fifths, and seven-twelfths. The value of unit fractions changes as the denominator changes because the numerator is always one. The unit fraction one-half requires two parts to make a whole. The unit fraction one-third requires three parts to make a whole. In the same way, if the unit fraction changes to one-thirteenth, then the number of parts required to make a whole becomes thirteen. An illustration of this is seen in the figure below. As the denominator increases, the size of the parts for each fraction decreases. As the bar goes from one-fourth to one-fifth, the size of the bars decreases, but the size of the denominator increases to five.

This pattern continues down the diagram; as the bars, or value of the fraction, get smaller, the denominator gets larger:

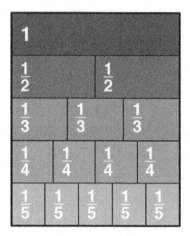

Using the Same Whole When Comparing Fractions

Comparing fractions requires the use of a common denominator. This necessity can be seen by the two pies below. The first pie has a shaded value of $\frac{2}{10}$ because two pieces are shaded out of the total of ten equal pieces. The second pie has a shaded value of $\frac{2}{7}$ because two pieces are shaded out of a total of seven equal pieces. These two fractions, two-tenths and two-sevenths, have the same numerator and so a misconception may be that they are equal. By looking at the shaded region in each pie, it is apparent that the fractions are not equal.

The numerators are the same, but the denominators are not. Two parts of a whole are not equivalent unless the whole is broken into the same number of parts. To compare the shaded regions, the denominators seven and ten must be made equal. The lowest number that the two denominators will both divide evenly into is 70, which is the lowest common denominator. Then the numerators must be converted by multiplying by the opposite denominator. These operations result in the two fractions $\frac{14}{70}$ and $\frac{20}{70}$. Now that these two have the same denominator, the conclusion can be made that $\frac{2}{7}$ represents a larger portion of the pie, as seen in the figure below:

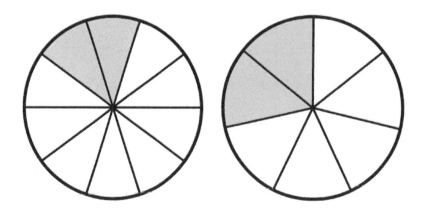

Order of Operations

As mentioned, the **order of operations** refers to the order in which problems are to be solved, from parenthesis or grouping, to addition and subtraction. A common way of remembering the order of operations is PEMDAS, or "Please Excuse My Dear Aunt Sally." The letters stand for parenthesis, or grouping, exponents, multiply/divide, and add/subtract.

The first step is to complete any operations inside the grouping symbols, or parenthesis. The next step is to simplify all exponents. After exponents, the operations of multiplication and division are performed in the order they appear from left to right. The last operations are addition and subtraction, also performed from left to right.

The following problem requires the use of order of operations to be solved:

$$2(3^3 + 5) - 8$$

The first step is to perform the operations inside the grouping symbols, or parenthesis. Inside the parenthesis, the exponent would be performed first, then the addition of $(3^3 + 5)$ which is $(27 + 5)$ or (32). These operations lead to the next step of $2(32) - 8$, where the multiplication can be performed between 2 and 8. This step leads to the problem $64 - 8$, where the answer is 56. The order of operations is important because if solved in a different order, the resulting number would not be 56. A common of when the order of operations can be used is when a store is having a sale and customers may use coupons. Other places may be at a restaurant, for the check, or the gas station when using a card to pay.

Representing Rational Numbers and Their Operations in Different Ways

Rational numbers can be written as fractions, but also as a percent or decimal. For example, three-fourths is a fraction written as three divided by four or $\frac{3}{4}$. It represents three parts out of a whole of four parts. By dividing three by four, the decimal of 0.75 is found. This decimal represents the same part to whole relationship as three-fourths. Seventy-five is in the hundredths place, so it can be read as 75 out of 100, the same ratio as 3 to 4. The decimal 0.75 is the same as 75 out of 100, or 75%. The rational number three-fourths represents the same portion as the decimal 0.75 and the percentage 75%.

Because there are different ways to represent rational numbers, the operations used to manipulate rational numbers can look different also. For the operation of multiplication, the problem can use a dot, an "x," or simply writing two variables side by side to indicate the need to find the product. Division can be represented by the line in a fraction or a division symbol. When adding or subtracting, the form of rational numbers is important and can be changed. Sometimes it is simpler to work with fractions, while sometimes decimals are easier to manipulate, depending on the operation. When comparing portion size, it may be easier to see each number as a percent, so it is important to understand the different ways rational numbers can be represented.

Representing Rational Numbers on a Number Line

A **number line** is a tool used to compare numbers by showing where they fall in relation to one another. Labeling a number line with integers is simple because they have no fractional component, and the values are easier to understand. The number line may start at −3 and go up to −2, then −1, then 0, and 1, 2, 3. This order shows that number 2 is larger than −1 because it falls further to the right on the number line.

When positioning rational numbers, the process may take more time because it requires that they all be in the same form. If they are percentages, fractions, and decimals, then conversions will have to be made to put them in the same form. For example, if the numbers $\frac{5}{4}$, 45%, and 2.38 need to be put in order on a number line, the numbers must first be transformed into one single form. Decimal form is an easy common ground because fractions can be changed by simply dividing, and percentages can be changed by moving the decimal point. After conversions are made, the list becomes 1.25, 0.45, and 2.38 respectively. Now the list is easier to arrange.

The number line with the list in order is shown in the top half of the graphic below in the order 0.45, 1.25, and 2.38:

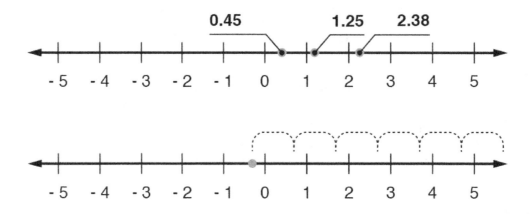

The sums and differences of rational numbers can be found using a number line after the rational numbers are put into the same form. This method is especially helpful when understanding the addition and subtraction of negative numbers. For example, the rational number six can be added to negative one-half using the number line. The following expression represents the problem: $-\frac{1}{2} + 6$. First, the original number $-\frac{1}{2}$ can be plotted by a dot on the number line, as shown in the lower half of the graphic above. Then 6 can be added by counting by whole numbers on the number line. The arcs on the graph represent the addition. The final answer is positive $5\frac{1}{2}$.

Illustrating Multiplication and Division Problems Using Equations, Rectangular Arrays, and Area Models

Multiplication and division can be represented by equations. These equations show the numbers involved and the operation. For example, "eight multiplied by six equals forty-eight" is seen in the following equation: $8 \times 6 = 48$. This operation can be modeled by rectangular arrays where one factor, 8, is the number of rows, and the other factor, 6, is the number of columns, as follows:

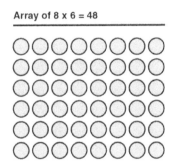

Rectangular arrays show what happens with the concept of multiplication. As one row of dots is drawn, that represents the first factor in the problem. Then the second factor is used to add the number of columns. The final model includes six rows of eight columns which results in forty-eight dots. These rectangular arrays show how multiplication of whole numbers will result in a number larger than the factors. Division can also be represented by equations and area models. A division problem such as "twenty-four divided by three equals eight" can be written as the following equation: $24 \div 8 = 3$. The object below shows an area model to represent the equation.

As seen in the model, the whole box represents 24 and the 3 sections represent the division by 3. In more detail, there could be 24 dots written in the whole box and each box could have 8 dots in it. Division shows how numbers can be divided into groups. For the example problem, it is asking how many numbers will be in each of the 3 groups that together make 24. The answer is 8 in each group.

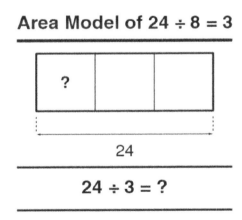

Practice Quiz

1. Joey and Sandy wanted to sell lemonade and cookies to earn some extra money. They sold cookies for $1 and lemonade for $0.50. At the end of the first day, they sold 24 cookies. The total money that they collected was $35.50. How many cups of lemonade did they sell?
   a. 11
   b. 21
   c. 22
   d. 23

2. Carla is starting a cake-decorating business and wants to know how long it will take her to start making a profit. She knows the original investment is $100. After that investment, she can begin making cakes and selling them for $20 each. How many cakes will she need to sell to break even on her investment?
   a. 5
   b. 100
   c. 10
   d. 2

3. If the ratio of x to y is 1:8, what is the product of x and y when $y = 48$?
   a. 6
   b. 183
   c. 98
   d. 288

4. The percent increase from 8 to 18 is equivalent to the percent increase from 234 to what number?
   a. 468
   b. 526
   c. 526.5
   d. 125

5. A jar is filled with green, yellow, and orange marbles. If $\frac{1}{4}$ of the marbles are green and $\frac{2}{7}$ are yellow, what fraction of the marbles are orange?
   a. $\frac{15}{28}$

   b. $\frac{13}{28}$

   c. $\frac{2}{3}$

   d. $\frac{3}{7}$

See answers on next page.

Answer Explanations

1. D: The following equation can be used to model how much money they collect for their sale: $M = 1c + 0.5l$, where M is money collected, c is cookies, and l is lemonade. By substituting 35.50 in for the variable M and 24 for the variable c, the equation can be solved for l, which is found to be 23 cups of lemonade. If there were 11 lemonades sold, the money collected would be $29.50. If 21 lemonades were sold, the money collected would be $34.50. If 22 lemonades were sold, the money collected would be $35.

2. A: The equation used to model this situation is $y = 20x - 100$, where 20 is price of each cake and 100 is the original investment. The value of x is the number of cakes and y is the money she makes. If this line is plotted on the graph, the x-intercept will be the number of cakes she needs to make to recoup her investment and break even. The x-intercept occurs when the y-value is zero. For this equation, setting $y = 0$ and solving for x gives a value of 5 cakes. 100 cakes would yield a profit of $1,900, ten cakes would yield a profit of $100, and 2 cakes would still leave her in the negative (-$10).

3. D: The ratio gives the proportion $\frac{x}{y} = \frac{1}{8}$. If $y = 48$, then $\frac{x}{48} = \frac{1}{8}$ means that $x = 6$. The product of x and y is therefore:

$$(6)(48) = 288$$

4. C: First, calculate the percent increase from 8 to 18 as:

$$\frac{18 - 8}{8} = 1.25 = 125\%$$

Then add 125% of 234 onto 234 to obtain:

$$292.5 + 234 = 526.5$$

5. B: The total fraction of green and yellow marbles is:

$$\frac{1}{4} + \frac{2}{7} = \frac{7}{28} + \frac{8}{28} = \frac{15}{28}$$

To find the fraction of marbles left (the orange marbles), subtract this value from 1.

$$1 - \frac{15}{28} = \frac{28}{28} - \frac{15}{28} = \frac{13}{28}$$

Word Knowledge

Word Knowledge

The Word Knowledge section of the ASVAB is designed to assess the test taker's vocabulary knowledge and skill in determining the answer choice with a meaning that most nearly matches that of the word presented in the question. In this way, Word Knowledge questions task test takers with applying their vocabulary skills to select the best synonyms.

Question Format

The questions in the Word Knowledge section are constructed very simply: the prompt is a single word, which is then followed by the five answer choices, each of which is also a single word. Test takers must consider the definition or meaning of the word provided in the prompt, and then select the answer choice that most nearly means the same thing. Consider the following example of a Word Knowledge question:

SERENE
   a. Serious
   b. Calm
   c. Tired
   d. Nervous

Test takers must consider the meaning of the given word (*serene*), read the four potential definitions or synonyms, and pick the word that most closely means the same thing as the word in the prompt. In this case, Choice *B* is the best answer because *serene* means calm. The format of this example question is exactly like all the 25 questions that will appear in the Word Knowledge section of the ASVAB.

Denotation and Connotation

Denotation, a word's explicit definition, is often set in comparison to connotation, the emotional, cultural, social, or personal implications associated with a word. Denotation is more of an objective definition, whereas connotation can be more subjective, although many connotative meanings of words are similar for certain cultures. The denotative meanings of words are usually based on facts, and the connotative meanings of words are usually based on emotion.

Here are some examples of words and their denotative and connotative meanings in Western culture:

Word	Denotative Meaning	Connotative Meaning
Home	A permanent place where one lives, usually as a member of a family.	A place of warmth; a place of familiarity; comforting; a place of safety and security. "Home" usually has a positive connotation.
Snake	A long reptile with no limbs and strong jaws that moves along the ground; some snakes have a poisonous bite.	An evil omen; a slithery creature (human or nonhuman) that is deceitful or unwelcome. "Snake" usually has a negative connotation.
Winter	A season of the year that is the coldest, usually from December to February in the northern hemisphere and from June to August in the southern hemisphere.	Circle of life, especially that of death and dying; cold or icy; dark and gloomy; hibernation, sleep, or rest. "Winter" can have a negative connotation, although many who have access to heat may enjoy the snowy season from their homes.

Analyzing Word Parts

By learning some of the etymologies of words and their parts, readers can break new words down into components and analyze their combined meanings. For example, the root word *soph* is Greek for wise or knowledge. Knowing this informs the meanings of English words including *sophomore, sophisticated,* and *philosophy.* Those who also know that *phil* is Greek for love will realize that *philosophy* means the love of knowledge. They can then extend this knowledge of *phil* to understand *philanthropist* (one who loves people), *bibliophile* (book lover), *philharmonic* (loving harmony), *hydrophilic* (water-loving), and so on. In addition, *phob-* derives from the Greek *phobos,* meaning fear. This informs all words ending with it as meaning fear of various things: *acrophobia* (fear of heights), *arachnophobia* (fear of spiders), *claustrophobia* (fear of enclosed spaces), *ergophobia* (fear of work), and *hydrophobia* (fear of water), among others.

Some words that originate from other languages, like ancient Greek, are found in a variety of English words. An advantage of the shared ancestry is that once readers recognize the meanings of Greek word roots, they can determine what many English words mean. As an example, the Greek word *métron* means to measure, a measure, or something used to measure; the English word meter derives from it. Knowing this informs many other English words, including *altimeter, barometer, diameter, hexameter, isometric,* and *metric.* While readers must know the meanings of the other parts of these words to decipher their meaning fully, they already have an idea that they are all related in some way to measures or measuring.

While all English words ultimately derive from a proto-language known as Indo-European, many of them historically came into the developing English vocabulary later, from sources like the ancient Greeks' language, the Latin used throughout Europe and much of the Middle East during the reign of the Roman Empire, and the Anglo-Saxon languages used by England's early tribes. In addition to classic revivals and native foundations, by the Renaissance era, other influences included French, German, Italian, and Spanish. Today we can often discern English word meanings by knowing common roots and affixes, particularly from Greek and Latin.

The following is a list of common prefixes and their meanings:

Prefix	Definition	Examples
a-	without	atheist, agnostic
ad-	to, toward	advance
ante-	before	antecedent, antedate
anti-	opposing	antipathy, antidote
auto-	self	autonomy, autobiography
bene-	well, good	benefit, benefactor
bi-	two	bisect, biennial
bio-	life	biology, biosphere
chron-	time	chronometer, synchronize
circum-	around	circumspect, circumference
com-	with, together	commotion, complicate
contra-	against, opposing	contradict, contravene
cred-	belief, trust	credible, credit
de-	from	depart
dem-	people	demographics, democracy
dis-	away, off, down, not	dissent, disappear
equi-	equal, equally	equivalent
ex-	former, out of	extract
for-	away, off, from	forget, forswear
fore-	before, previous	foretell, forefathers
homo-	same, equal	homogenized
hyper-	excessive, over	hypercritical, hypertension
in-	in, into	intrude, invade
inter-	among, between	intercede, interrupt
mal-	bad, poorly, not	malfunction
micr-	small	microbe, microscope
mis-	bad, poorly, not	misspell, misfire
mono-	one, single	monogamy, monologue
mor-	die, death	mortality, mortuary
neo-	new	neolithic, neoconservative
non-	not	nonentity, nonsense
omni-	all, everywhere	omniscient
over-	above	overbearing
pan-	all, entire	panorama, pandemonium
para-	beside, beyond	parallel, paradox
phil-	love, affection	philosophy, philanthropic
poly-	many	polymorphous, polygamous
pre-	before, previous	prevent, preclude

Prefix	Definition	Examples
prim-	first, early	primitive, primary
pro-	forward, in place of	propel, pronoun
re-	back, backward, again	revoke, recur
sub-	under, beneath	subjugate, substitute
super-	above, extra	supersede, supernumerary
trans-	across, beyond, over	transact, transport
ultra-	beyond, excessively	ultramodern, ultrasonic, ultraviolet
un-	not, reverse of	unhappy, unlock
vis-	to see	visage, visible

The following is a list of common suffixes and their meanings:

Suffix	Definition	Examples
-able	likely, able to	capable, tolerable
-ance	act, condition	acceptance, vigilance
-ard	one that does excessively	drunkard, wizard
-ation	action, state	occupation, starvation
-cy	state, condition	accuracy, captaincy
-er	one who does	teacher
-esce	become, grow, continue	convalesce, acquiesce
-esque	in the style of, like	picturesque, grotesque
-ess	feminine	waitress, lioness
-ful	full of, marked by	thankful, zestful
-ible	able, fit	edible, possible, divisible
-ion	action, result, state	union, fusion
-ish	suggesting, like	churlish, childish
-ism	act, manner, doctrine	barbarism, socialism
-ist	doer, believer	monopolist, socialist
-ition	action, result, state,	sedition, expedition
-ity	quality, condition	acidity, civility
-ize	cause to be, treat with	sterilize, mechanize, criticize
-less	lacking, without	hopeless, countless
-like	like, similar	childlike, dreamlike
-ly	like, of the nature of	friendly, positively
-ment	means, result, action	refreshment, disappointment
-ness	quality, state	greatness, tallness
-or	doer, office, action	juror, elevator, honor
-ous	marked by, given to	religious, riotous
-some	apt to, showing	tiresome, lonesome
-th	act, state, quality	warmth, width
-ty	quality, state	enmity, activity

The following is a list of root words and their meanings:

Root	Definition	Examples
ambi	both	ambidextrous, ambiguous
anthropo	man; humanity	anthropomorphism, anthropology
auto	self	automobile, autonomous
bene	good	benevolent, benefactor
bio	life	biology, biography
chron	time	chronology
circum	around	circumvent, circumference
dyna	power	dynasty, dynamite
fort	strength	fortuitous, fortress
graph	writing	graphic
hetero	different	heterogeneous
homo	same	homonym, homogenous
hypo	below, beneath	hypothermia
morph	shape; form	morphology
mort	death	mortal, mortician
multi	many	multimedia, multiplication
nym	name	antonym, synonym
phobia	fear	claustrophobia
port	carry	transport
pseudo	false	pseudoscience, pseudonym
scope	viewing instrument	telescope, microscope
techno	art; science; skill	technology, techno
therm	heat	thermometer, thermal
trans	across	transatlantic, transmit
under	too little	underestimate

Practice Quiz

1. Surly most nearly means
   a. Amiable
   b. Irritable
   c. Fatigued
   d. Clever

2. Savvy most nearly means
   a. Foolish
   b. Handsome
   c. Formidable
   d. Smart

3. Clout most nearly means
   a. Disposition
   b. Uncanny
   c. Prestige
   d. Sympathy

4. Petulant most nearly means
   a. Cranky
   b. Pleasant
   c. Sensible
   d. Greedy

5. Nuance most nearly means
   a. Reform
   b. Subtlety
   c. Pinnacle
   d. Motion

See answers on next page.

Answer Explanations

1. B: *Surly* most nearly means irritable. *Surly* means agitated or quick to anger. Choice *A*, *amiable*, means friendly, so this is incorrect. Choice *C* means tired, so this is incorrect. *Clever* means funny or witty, so this is incorrect.

2. D: *Savvy* means someone is intelligent or knowledgeable, so its synonym is Choice *D*, *smart*. Choice *A*, foolish, is an antonym for *savvy*. Handsome means good-looking and does not apply to being savvy. *Formidable*, Choice *C*, means dangerous, so this is incorrect.

3. C: *Clout* means to have prestige or influence, so Choice *C* is the best answer. *Disposition* refers to someone's personal temperament, so Choice *A* is incorrect. *Uncanny* means strange, so Choice *B* is incorrect. *Sympathy* is a shared feeling that denotes compassion, so Choice *D* is incorrect.

4. A: *Petulant* means cranky or whiny, so Choice *A* is the best answer. *Pleasant* is incorrect, as it is an antonym for petulant. *Sensible* means to be rational and realistic, so Choice *C* is incorrect. Although *greedy* denotes a negative connotation like *petulant*, it means selfish, not cranky. Therefore, Choice *D* is incorrect.

5. B: *Nuance* most nearly means a slight difference or subtlety in something, so Choice *B* is the best choice. *Reform* means a change or correction, so Choice *A* is incorrect. *Pinnacle* means the highest level, so Choice *C* is incorrect. *Motion* denotes movement or activity, so Choice *D* is incorrect.

Paragraph Comprehension

Inferring the Logical Conclusion from a Reading Selection

An inference is an educated guess or conclusion based on sound evidence and reasoning within the text. The test may include multiple-choice questions asking about the logical conclusion that can be drawn from reading a text, and you will have to identify the choice that unavoidably leads to that conclusion. In order to eliminate the incorrect choices, the test-taker should come up with a hypothetical situation wherein an answer choice is true, but the conclusion is not true. Here is an example:

> Fred purchased the newest PC available on the market. Therefore, he purchased the most expensive PC in the computer store.
> What can one assume for this conclusion to follow logically?

> a. Fred enjoys purchasing expensive items.
> b. PCs are some of the most expensive personal technology products available.
> c. The newest PC is the most expensive one.

The premise of the text is the first sentence: Fred purchased the newest PC. The conclusion is the second sentence: Fred purchased the most expensive PC. Recent release and price are two different factors; the difference between them is the logical gap. To eliminate the gap, one must connect the new information from the conclusion with the pertinent information from the premise. In this example, there must be a connection between product recency and product price. Therefore, a possible bridge to the logical gap could be a sentence stating that the newest PCs always cost the most.

Identifying the Topic, Main Idea, and Supporting Details

The **topic** of a text is the general subject matter. Text topics can usually be expressed in one word, or a few words at most. Additionally, readers should ask themselves what point the author is trying to make. This point is the **main idea** of the text, the one thing the author wants readers to know concerning the topic. Once the author has established the main idea, they will support the main idea by supporting details. **Supporting details** are evidence that support the main idea and include personal testimonies, examples, or statistics.

One analogy for these components and their relationships is that a text is like a well-designed house. The topic is the roof, covering all rooms. The main idea is the frame. The supporting details are the various rooms. To identify the topic of a text, readers can ask themselves what or who the author is writing about in the paragraph. To locate the main idea, readers can ask themselves what one idea the author wants readers to know about the topic. To identify supporting details, readers can put the main idea into question form and ask, "what does the author use to prove or explain their main idea?"

Let's look at an example. An author is writing an essay about the Amazon rainforest and trying to convince the audience that more funding should go into protecting the area from deforestation. The author makes the argument stronger by including evidence of the benefits of the rainforest: it provides habitats to a variety of species, it provides much of the earth's oxygen which in turn cleans the atmosphere, and it is the home to medicinal plants that may be the answer to some of the world's deadliest diseases.

Here is an outline of the essay looking at topic, main idea, and supporting details:

Topic: Amazon rainforest
Main Idea: The Amazon rainforest should receive more funding in order to protect it from deforestation.
Supporting Details:
1. It provides habitats to a variety of species
2. It provides much of the earth's oxygen which in turn cleans the atmosphere
3. It is home to medicinal plants that may be the answer to some of the deadliest diseases.

Notice that the topic of the essay is listed in a few key words: "Amazon rainforest." The main idea tells us what about the topic is important: that the topic should be funded in order to prevent deforestation. Finally, the supporting details are what author relies on to convince the audience to act or to believe in the truth of the main idea.

Distinguishing Between Fact and Opinion, Biases, and Stereotypes

Facts and Opinions

A **fact** is a statement that is true empirically or an event that has actually occurred in reality and can be proven or supported by evidence; it is generally objective. In contrast, an **opinion** is subjective, representing something that someone believes rather than something that exists in the absolute. People's individual understandings, feelings, and perspectives contribute to variations in opinion. Though facts are typically objective in nature, in some instances, a statement of fact may be both factual and yet also subjective. For example, emotions are individual subjective experiences. If an individual says that they feel happy or sad, the feeling is subjective, but the statement is factual; hence, it is a subjective fact. In contrast, if one person tells another that the other is feeling happy or sad—whether this is true or not— that is an assumption or an opinion.

Biases

Biases usually occur when someone allows their personal preferences or ideologies to interfere with what should be an objective decision. In personal situations, someone is biased towards someone if they favor them in an unfair way. In academic writing, being biased in your sources means leaving out objective information that would turn the argument one way or the other. The evidence of bias in academic writing makes the text less credible, so be sure to present all viewpoints when writing, not just your own, so to avoid coming off as biased. Being objective when presenting information or dealing with people usually allows the person to gain more credibility.

Stereotypes

Stereotypes are preconceived notions that place a particular rule or characteristics on an entire group of people. Stereotypes are usually offensive to the group they refer to or allies of that group and often have negative connotations. The reinforcement of stereotypes isn't always obvious. Sometimes stereotypes can be very subtle and are still widely used for people to understand categories within the world. For example, saying that women are more emotional and intuitive than men is a stereotype, although this is still an assumption used by many to understand the differences between one another.

Interpreting the Meaning of Words and Phrases Using Context

When readers encounter an unfamiliar word in text, they can use the surrounding **context**—the overall subject matter, specific chapter/section topic, and especially the immediate sentence context. Among others, one category of context clues is grammar. For example, the position of a word in a sentence and its relationship to the other words can help the reader establish whether the unfamiliar word is a verb, a noun, an adjective, an adverb, etc. This narrows down the possible meanings of the word to one part of speech. However, this may be insufficient. In a sentence that many birds *migrate* twice yearly, the reader can determine the word is a verb, and probably does not mean eat or drink; but it could mean travel, mate, lay eggs, hatch, molt, etc.

Some words can have a number of different meanings depending on how they are used. For example, the word *fly* has a different meaning in each of the following sentences:

- "His trousers have a fly on them."
- "He swatted the fly on his trousers."
- "Those are some fly trousers."
- "They went fly fishing."
- "She hates to fly."
- "If humans were meant to fly, they would have wings."

As strategies, readers can try substituting a familiar word for an unfamiliar one and see whether it makes sense in the sentence. They can also identify other words in a sentence, offering clues to an unfamiliar word's meaning.

Identifying Primary Sources in Various Media

A **primary source** is a piece of original work, which can include books, musical compositions, recordings, movies, works of visual art (paintings, drawings, photographs), jewelry, pottery, clothing, furniture, and other artifacts. Within books, primary sources may be of any genre. Whether nonfiction based on actual events or a fictional creation, the primary source relates the author's firsthand view of some specific event, phenomenon, character, place, process, ideas, field of study or discipline, or other subject matter. Whereas primary sources are original treatments of their subjects, secondary sources are a step removed from the original subjects; they analyze and interpret primary sources. These include journal articles, newspaper or magazine articles, works of literary criticism, political commentaries, and academic textbooks.

In the field of history, primary sources frequently include documents that were created around the same time period that they were describing, and most often produced by someone who had direct experience or knowledge of the subject matter. In contrast, secondary sources present the ideas and viewpoints of other authors about the primary sources; in history, for example, these can include books and other written works about the particular historical periods or eras in which the primary sources were produced. Primary sources pertinent in history include diaries, letters, statistics, government information, and original journal articles and books. In literature, a primary source might be a literary novel, a poem or book of poems, or a play. Secondary sources addressing primary sources may be criticism, dissertations, theses, and journal articles. Tertiary sources, typically reference works referring to primary and secondary sources, include encyclopedias, bibliographies, handbooks, abstracts, and periodical indexes.

In scientific fields, when scientists conduct laboratory experiments to answer specific research questions and test hypotheses, lab reports and reports of research results constitute examples of primary sources. When researchers produce statistics to support or refute hypotheses, those statistics are primary sources. When a scientist is studying some subject longitudinally or conducting a case study, they may keep a journal or diary. For example, Charles Darwin kept diaries of extensive notes on his studies during sea voyages on the *Beagle*, visits to the Galápagos Islands, etc.; Jean Piaget kept journals of observational notes for case studies of children's learning behaviors. Many scientists, particularly in past centuries, shared and discussed discoveries, questions, and ideas with colleagues through letters, which also constitute primary sources. When a scientist seeks to replicate another's experiment, the reported results, analysis, and commentary on the original work is a secondary source, as is a student's dissertation if it analyzes or discusses others' work rather than reporting original research or ideas.

Practice Quiz

1. He looked down and said in a condescending tone, "One day, lad, when you are a bit older, a bit more experienced with the workings of this world, you will make a fine citizen, like myself." Everyone listening rolled their eyes at him and thought to themselves that he is as arrogant as one can be.

Why did the people think this man was arrogant?
   a. He was a fine citizen
   b. He was experienced and they were not
   c. He thought too highly of himself
   d. He called someone a "lad"

2. The list seemed to never stop. He was an award-winning scientist who held three doctorate level degrees. He was rich beyond anyone's dreams, he was handsome, he even had a good sense of humor. Yet, despite this, he was empty inside. None of these things made him happy. All he could do was think about what life would have been like had he never known fame or success.

Why is this man not happy?
   a. He lost a loved one
   b. He did not have any friends
   c. Money cannot actually make someone happy
   d. Not enough information to know

3. What an exciting day! We get to explore a strange cave filled with treasure, climb a giant mountain covered in snow, and eat food from eight different countries! All of this and we do not even need to leave our bedroom. Who needs to travel when you got kids to play with?

What is most likely true based on this passage?
   a. This person is going to an exotic location
   b. This person is playing a pretend game
   c. This person is mentally impaired
   d. This person is daydreaming

4. Tick, tock, tick, tock. The sound nearly drove a man mad. Was there nothing else to listen to? Tick. Was there no one to speak with? Tock. There is not even a cough or a sneeze that can be heard. Tick. How much longer until this torture ends? Tock. Finally, after what felt like forever, the bell rang, and the children turned in their exams and went home.

What location did this passage describe?
   a. School
   b. Prison
   c. An insane asylum
   d. A boarding school

5. The waves crashed hard against the side of the boat. There was not a lot of rain, but the wind and the waves caused blue-green sea water to spray on everyone. The captain gave the order for all hands-on deck. This was going to be a tough storm to beat. Everyone ran to their stations and began busying themselves with tying and untying various things or scooping out water with large buckets.

What does the phrase "all hands-on deck" mean?
   a. Everyone needs to clean the deck
   b. Some people need to go to the deck to work
   c. Everyone needs to go to the deck to work
   d. There is a bad storm coming

See answers on next page.

Answer Explanations

1. C: This man in the story clearly thought much of himself since he called himself a fine citizen and spoke in condescending tones to someone else about it. Choice *A* is unknown since the man could have been a fine citizen, but that would not make him appear arrogant. Choice *B* does not necessitate arrogance either since people can be experienced and not arrogant. Choice *D* does not, in and of itself, make someone arrogant because the word *lad* simply refers to a young boy.

2. D: The information given in this passage is not sufficient to know the source of this man's discontentment. There is no indication as to whether he has friends, has lost a loved one, or if his money is causing his sadness.

3. B: Based on the information given in the passage, it is clear that this person is playing a pretend game with their kids. Since the text ends with saying, "We do not even need to leave our bedroom," it is clear that they are not traveling to an exotic place, Choice *A*. Nothing in the text indicates any mental problems with this person, Choice *C*. Finally, since the passage says, "Who needs to travel when you have kids," it is evident that this person is not day dreaming but involved with children. Therefore, playing a pretend game is most likely.

4. A: This passage begins with describing a very quiet place that someone would be eager to leave. The tone of the passage makes it sound as though this could be a prison or an insane asylum. However, by the end of the text, it is evident that the passage is about a school since children are turning in exams and then going home. Since the children get to actually go home, this could not be a boarding school.

5. C: Based on the context of this passage, Choice *C* is the most likely answer. Choice *A*, everyone needs to clean the deck, is not right because the story clearly indicates something bad is happening and would not be a good time to clean. Choice *B*, some people need to go to the deck to work, is refuted by the phrase itself being "all hands-on deck" as opposed to "some hands-on deck." Finally, Choice *D*, there is a bad storm coming, may in fact be true, but since the passage says that everyone ran to the deck and started doing various things shows that the command referred to action that people needed to take.

Mathematics Knowledge

Exponents and Roots

The nth root of a is given as $\sqrt[n]{a}$, which is called a **radical**. Typical values for n are 2 and 3, which represent the square and cube roots. In this form, n represents an integer greater than or equal to 2, and a is a real number. If n is even, a must be nonnegative, and if n is odd, a can be any real number. This radical can be written in exponential form as $a^{\frac{1}{n}}$. Therefore, $\sqrt[4]{15}$ is the same as $15^{\frac{1}{4}}$ and $\sqrt[3]{-5}$ is the same as $(-5)^{\frac{1}{3}}$.

In a similar fashion, the nth root of a can be raised to a power m, which is written as $\left(\sqrt[n]{a}\right)^m$. This expression is the same as $\sqrt[n]{a^m}$. For example,

$$\sqrt[2]{4^3} = \sqrt[2]{64} = 8 = \left(\sqrt[2]{4}\right)^3 = 2^3$$

Because $\sqrt[n]{a} = a^{\frac{1}{n}}$, both sides can be raised to an exponent of m, resulting in:

$$\left(\sqrt[n]{a}\right)^m = \sqrt[n]{a^m} = a^{\frac{m}{n}}$$

This rule allows:

$$\sqrt[2]{4^3} = \left(\sqrt[2]{4}\right)^3 = 4^{\frac{3}{2}}$$

$$(2^2)^{\frac{3}{2}} = 2^{\frac{6}{2}} = 2^3 = 8$$

Negative exponents can also be incorporated into these rules. Any time an exponent is negative, the base expression must be flipped to the other side of the fraction bar and rewritten with a positive exponent. For instance,

$$2^{-3} = \frac{1}{2^3} = \frac{1}{8}$$

Therefore, two more relationships between radical and exponential expressions are:

$$a^{-\frac{1}{n}} = \frac{1}{\sqrt[n]{a}}$$

$$a^{-\frac{m}{n}} = \frac{1}{\sqrt[n]{a^m}} = \frac{1}{\left(\sqrt[n]{a}\right)^m}$$

Thus,

$$8^{-\frac{1}{3}} = \frac{1}{\sqrt[3]{8}} = \frac{1}{2}$$

All these relationships are very useful when simplifying complicated radical and exponential expressions. If an expression contains both forms, use one of these rules to change the expression to contain either all radicals or all exponential expressions. This process makes the entire expression much easier to work with, especially if the expressions are contained within equations.

Consider the following example:

$$\sqrt{x} \times \sqrt[4]{x}$$

It is written in radical form; however, it can be simplified into one radical by using exponential expressions first. The expression can be written as:

$$x^{\frac{1}{2}} \times x^{\frac{1}{4}}$$

It can be combined into one base by adding the exponents as:

$$x^{\frac{1}{2}+\frac{1}{4}} = x^{\frac{3}{4}}$$

Writing this back in radical form, the result is $\sqrt[4]{x^3}$.

Using Structure to Isolate or Identify a Quantity of Interest

When solving equations, it is important to note which quantity must be solved for. This quantity can be referred to as the **quantity of interest**. The goal of solving is to isolate the variable in the equation using logical mathematical steps. The **addition property of equality** states that the same real number can be added to both sides of an equation and equality is maintained. Also, the same real number can be subtracted from both sides of an equation to maintain equality. Second, the **multiplication property of equality** states that the same nonzero real number can multiply both sides of an equation, and still, equality is maintained. Because division is the same as multiplying times a reciprocal, an equation can be divided by the same number on both sides as well.

When solving inequalities, the same ideas are used. However, when multiplying by a negative number on both sides of an inequality, the inequality symbol must be flipped in order to maintain the logic. The same is true when dividing both sides of an inequality by a negative number.

Basically, in order to isolate a quantity of interest in either an equation or inequality, the same thing must be done to both sides of the equals sign, or inequality symbol, to keep everything mathematically correct.

Interpreting the Variables and Constants in Expressions

A linear function of the form $f(x) = mx + b$ has two important quantities: m and b. The quantity m represents the slope of the line, and the quantity b represents the y-intercept of the line. When the function represents a real-life situation or a mathematical model, these two quantities are very meaningful. The **slope**, m, represents the rate of change, or the amount y increases or decreases given an increase in x. If m is positive, the rate of change is positive, and if m is negative, the rate of change is negative. The y-intercept, b, represents the amount of quantity y when x is 0. In many applications, if the x-variable is never a negative quantity, the y-intercept represents the initial amount of the quantity y. The x-variable often represents time, so it makes sense that it would not be negative.

Consider the following example. These two equations represent the cost, C, of t-shirts, x, at two different printing companies:

$$C(x) = 7x$$

$$C(x) = 5x + 25$$

The first equation represents a scenario in which each t-shirt costs $7. In this equation, x varies directly with y. There is no y-intercept, which means that there is no initial cost for using that printing company. The rate of change is 7, which is price per shirt. The second equation represents a scenario that has both an initial cost and a cost per t-shirt. The slope of 5 shows that each shirt is $5. The y-intercept of 25 shows that there is an initial cost of using that company. Therefore, it makes sense to use the first company at $7 per shirt when only purchasing a small number of t-shirts. However, any large orders would be cheaper from the second company because eventually that initial cost would become negligible.

Recognizing and Representing Patterns

Number and Shape Patterns

Patterns in math are those sets of numbers or shapes that follow a rule. Given a set of values, patterns allow the question of "what's next?" to be answered. In the following set, there are two types of shapes, a white rectangle and a gray circle. The set contains a pattern because every odd-placed shape is a white rectangle, and every even-placed spot is taken by a gray circle. This is a pattern because there is a rule of white rectangle, then gray circle, that is followed to find the set.

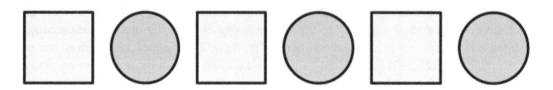

A set of numbers can also be described as having a pattern if there is a rule that can be followed to reproduce the set. The following set of numbers has a rule of adding 3 each time. It begins with zero and increases by 3 each time. By following this rule and pattern, the number after 12 is found to be 15. Further extending the pattern, the numbers are 18, 21, 24, 27. The pattern of increasing by multiples of three can describe this pattern.

A pattern can also be generated from a given rule. Starting with zero, the rule of adding 5 can be used to produce a set of numbers. The following list will result from using the rule: 0, 5, 10, 15, 20. Describing this pattern can include words such as "multiples" of 5 and an "increase" of 5. Any time this pattern needs to be extended, the rule can be applied to find more numbers. Patterns are identified by the rules they follow. This rule should be able to generate new numbers or shapes, while also applying to the given numbers or shapes.

Making Predictions Based on Patterns

Given a certain pattern, future numbers or shapes can be found. Pascal's triangle is an example of a pattern of numbers. Questions can be asked of the triangle, such as, "what comes next?" and "what values determine the next line?" By examining the different parts of the triangle, conjectures can be made about how the numbers are generated. For the first few rows of numbers, the increase is small. Then the

102

numbers begin to increase more quickly. By looking at each row, a conjecture can be made that the sum of the first row determines the second row's numbers. The second row's numbers can be added to find the third row. To test this conjecture, two numbers can be added, and the number found directly between and below them should be that sum. For the third row, the middle number is 2, which is the sum of the two 1s above it. For the fifth row, the 1 and 3 can be added to find a sum of 4, the same number below the 1 and 3. This pattern continues throughout the triangle.

Once the pattern is confirmed to apply throughout the triangle, predictions can be made for future rows. The sums of the bottom row numbers can be found and then added to the bottom of the triangle. In more general terms, the diagonal rows have patterns as well. The outside numbers are always 1. The second diagonal rows are in counting order. The third diagonal row increases each time by one more than the previous. It is helpful to generalize patterns because it makes the pattern more useful in terms of applying it. Pascal's triangle can be used to predict the tossing of a coin, predicting the chances of heads or tails for different numbers of tosses.

It can also be used to show the Fibonacci Sequence, which is found by adding the diagonal numbers together.

Pascal's Triangle

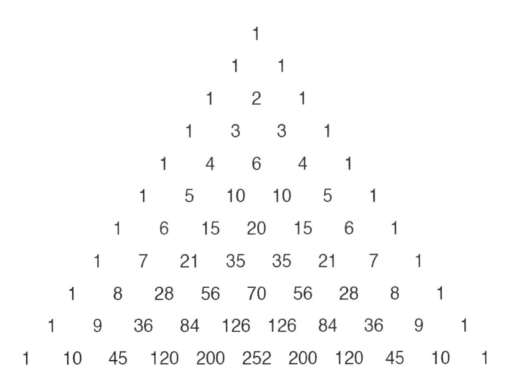

Relationships Between the Corresponding Terms of Two Numerical Patterns

Sets of numerical patterns can be found by starting with a number and following a given rule. If two sets are generated, the corresponding terms in each set can be found to relate to one another by one or more operations. For example, the following table shows two sets of numbers that each follow their own pattern. The first column shows a pattern of numbers increasing by 1. The second column shows the numbers increasing by 4. The numbers in the first column correspond to those in the second. A question to ask is, "How can the number in the first column turn into the number in the second column?"

1	4
2	8
3	12
4	16
5	20

This answer will lead to the relationship between the two sets. By recognizing the multiples of 4 in the right column and the counting numbers in order in the left column, the relationship of multiplying by four is determined. The first set is multiplied by 4 to get the second set of numbers. To confirm this relationship, each set of corresponding numbers can be checked. For any two sets of numerical patterns, the corresponding numbers can be lined up to find how each one relates to the other. In some cases, the relationship is simply addition or subtraction, multiplication or division. In other relationships, these operations are used in conjunction with each together. The relationship in the following table uses both multiplication and addition. The following expression shows this relationship: $3x + 2$. The x represents the numbers in the first column.

1	5
2	8
3	11
4	14

Algebra

Algebraic Expressions Versus Equations

An **algebraic expression** is a mathematical phrase that may contain numbers, variables, and mathematical operations. An expression represents a single quantity. For example, $3x + 2$ is an algebraic expression.

An **algebraic equation** is a mathematical sentence with two expressions that are equal to each other. That is, an equation must contain an equals sign, as in:

$$3x + 2 = 17$$

This statement says that the value of the expression on the left side of the equals sign is equivalent to the value of the expression on the right side. In an expression, there are not two sides because there is no equals sign. The equals sign ($=$) is the difference between an expression and an equation.

To distinguish an expression from an equation, just look for the equals sign.

Example: Determine whether each of these is an expression or an equation.

    a. $16 + 4x = 9x - 7$ Solution: Equation

    b. $-27x - 42 + 19y$ Solution: Expression

    c. $4 = x + 3$ Solution: Equation

Adding and Subtracting Linear Algebraic Expressions

To add and subtract linear algebraic expressions, you must combine like terms. **Like terms** are terms that have the same variable with the same exponent. In the following example, the x-terms can be added because the variable and exponent are the same. These terms add to be $9x$. Terms without a variable component are called **constants**. These terms will add to be nine.

Example: Add $(3x - 5) + (6x + 14)$

        $3x - 5 + 6x + 14$ Rewrite without parentheses

        $3x + 6x - 5 + 14$ Use the commutative property of addition

        $9x + 9$ Combine like terms

When subtracting linear expressions, be careful to add the opposite when combining like terms. Do this by distributing –1, which is multiplying each term inside the second parenthesis by negative one. Remember that distributing –1 changes the sign of each term.

Example: Subtract $(17x + 3) - (27x - 8)$

        $17x + 3 - 27x + 8$ Use the distributive property

        $17x - 27x + 3 + 8$ Use the commutative property of addition

        $-10x + 11$ Combine like terms

Example: Simplify by adding or subtracting:

   $(6m + 28z - 9) + (14m + 13) - (-4z + 8m + 12)$

   $6m + 28z - 9 + 14m + 13 + 4z - 8m - 12$ Use the distributive property

   $6m + 14m - 8m + 28z + 4z - 9 + 13 - 12$ Use the commutative property of addition

   $12m + 32z - 8$ Combine like terms

Using the Distributive Property to Generate Equivalent Linear Algebraic Expressions

The Distributive Property:

$$a(b + c) = ab + ac$$

The **distributive property** is a way of taking a factor and multiplying it through a given expression in parentheses. Each term inside the parentheses is multiplied by the outside factor, eliminating the parentheses. The following example shows how to distribute the number 3 to all the terms inside the parentheses.

Example: Use the distributive property to write an equivalent algebraic expression:

$$3(2x + 7y + 6)$$

$$3(2x) + 3(7y) + 3(6) \qquad \text{Use the distributive property}$$

$$6x + 21y + 18 \qquad \text{Simplify}$$

Because $a - b$ can be written $a + (-b)$, the distributive property can be applied in the example below.

Example: Use the distributive property to write an equivalent algebraic expression.

$$7(5m - 8)$$

$$7[5m + (-8)] \qquad \text{Rewrite subtraction as addition of –8}$$

$$7(5m) + 7(-8) \qquad \text{Use the distributive property}$$

$$35m - 56 \qquad \text{Simplify}$$

In the following example, note that the factor of 2 is written to the right of the parentheses but is still distributed as before:

Example: Use the distributive property to write an equivalent algebraic expression:

$$(3m + 4x - 10)2$$

$$(3m)2 + (4x)2 + (-10)2 \qquad \text{Use the distributive property}$$

$$6m + 8x - 20 \qquad \text{Simplify}$$

Example: $-(-2m + 6x)$

In this example, the negative sign in front of the parentheses can be interpreted as $-1(-2m + 6x)$

$$1(-2m + 6x)$$

$$-1(-2m) + (-1)(6x) \qquad \text{Use distributive property}$$

$$2m - 6x \qquad \text{Simplify}$$

Evaluating Simple Algebraic Expressions for Given Values of Variables

To evaluate an algebra expression for a given value of a variable, replace the variable with the given value. Then perform the given operations to simplify the expression.

Example: Evaluate $12 + x$ for $x = 9$

$12 + (9)$	Replace x with the value of 9 as given in the problem. It is a good idea to always use parentheses when substituting this value. This will be particularly important in the following examples.
21	Add

Now see that when x is 9, the value of the given expression is 21.

Example: Evaluate $4x + 7$ for $x = 3$

$4(3) + 7$	Replace the x in the expression with 3
$12 + 7$	Multiply (remember order of operations)
19	Add

Therefore, when x is 3, the value of the given expression is 19.

Example: Evaluate $-7m - 3r - 18$ for $m = 2$ and $r = -1$

$-7(2) - 3(-1) - 18$	Replace m with 2 and r with –1
$-14 + 3 - 18$	Multiply
-29	Add

So, when m is 2 and r is –1, the value of the given expression is –29.

Mathematical Terms That Identify Parts of Expressions

A **variable** is a symbol used to represent a number. Letters, like x, y, and z, are often used as variables in algebra.

A **constant** is a number that cannot change its value. For example, 18 is a constant.

A **term** is a constant, variable, or the product of constants and variables. In an expression, terms are separated by $+$ and $-$ signs. Examples of terms are $24x$, -32, and $15xyz$.

Like terms are terms that contain the same variables in the same powers. For example, $6z$ and $-8z$ are like terms, and $9xy$ and $17xy$ are like terms. $12y^3$ and $3y^3$ are also like terms. Lastly, constants, like 23 and 51, are like terms as well.

A **factor** is something that is multiplied by something else. A factor may be a constant, a variable, or a sum of constants or variables.

A **coefficient** is the numerical factor in a term that has a variable. In the term $16x$, the coefficient is 16.

Example: Given the below expression, answer the following questions:

$$6x - 12y + 18$$

1. How many terms are in the expression?

    Solution: 3

2. Name the terms.

    Solution: $6x$, $-12y$, and 18 (Notice that the minus sign preceding the 12 is interpreted to represent negative 12)

3. Name the factors.

    Solution: 6, x, -12, y

4. What are the coefficients in this expression?

    Solution: 6 and -12

5. What is the constant in this expression?

    Solution: 18

Using Formulas to Determine Unknown Quantities

Given the formula for the area of a rectangle, $A = lw$, with $A = area$, $l = length$, and $w = width$, the area of a rectangle can be determined, given the length and the width.

For example, if the length of a rectangle is 7 cm and the width is 10 cm, find the area of the rectangle. Just as when evaluating expressions, to solve, replace the variables with the given values. Thus, given $A = lw$, and $l = 7$ and $w = 10$, $A = (7)(10)$, which equals 70. Therefore, the area of the rectangle is 70 cm².

Consider an example using the formula for perimeter of a rectangle, which is $P = 2l + 2w$, where P is perimeter, l is length, and w is width. If the length of a rectangle is 12 inches and the width is 9 inches, find the perimeter.

Solution: $P = 2l + 2w$

$P = 2(12) + 2(9)$	Replace l with 12 and w with 9
$P = 24 + 18$	Use correct order of operations; multiply first
$P = 42$	Add

The perimeter of this rectangle is 42 inches.

Solving Real-World One- or Multi-Step Problems with Rational Numbers

One-step problems take only one mathematical step to solve. For example, solving the equation $5x = 45$ is a one-step problem because the one step of dividing both sides of the equation by 5 is the only step necessary to obtain the solution $x = 9$. The multiplication principle of equality is the one step used to isolate the variable. The equation is of the form $ax = b$, where a and b are rational numbers. Similarly, the addition principle of equality could be the one step needed to solve a problem. In this case, the equation would be of the form $x + a = b$ or $x - a = b$, for real numbers a and b.

A multi-step problem involves more than one step to find the solution, or it could consist of solving more than one equation. An equation that involves both the addition principle and the multiplication principle is a two-step problem, and an example of such an equation is $2x - 4 = 5$. To solve, add 4 to both sides and then divide both sides by 2. An example of a two-step problem involving two separate equations is:

$$y = 3x$$

$$2x + y = 4$$

The two equations form a system that must be solved together in two variables. The system can be solved by the substitution method. Since y is already solved for in terms of x, replace y with $3x$ in the equation $2x + y = 4$, resulting in:

$$2x + 3x = 4$$

Therefore, $5x = 4$ and $x = \frac{4}{5}$. Because there are two variables, the solution consists of a value for both x and for y. Substitute $x = \frac{4}{5}$ into either original equation to find y. The easiest choice is $y = 3x$. Therefore:

$$y = 3 \times \frac{4}{5} = \frac{12}{5}$$

The solution can be written as the ordered pair $\left(\frac{4}{5}, \frac{12}{5}\right)$.

Real-world problems can be translated into both one-step and multi-step problems. In either case, the word problem must be translated from the verbal form into mathematical expressions and equations that can be solved using algebra. An example of a one-step real-world problem is the following: A cat weighs half as much as a dog living in the same house. If the dog weighs 14.5 pounds, how much does the cat weigh? To solve this problem, an equation can be used. In any word problem, the first step is to define variables that represent the unknown quantities. For this problem, let x be equal to the unknown weight of the cat. Because two times the weight of the cat equals 14.5 pounds, the equation to be solved is: $2x = 14.5$. Use the multiplication principle to divide both sides by 2. Therefore, $x = 7.25$, and the cat weighs 7.25 pounds.

Most of the time, real-world problems require multiple steps. The following is an example of a multi-step problem: The sum of two consecutive page numbers is equal to 437. What are those page numbers? First, define the unknown quantities. If x is equal to the first page number, then $x + 1$ is equal to the next page number because they are consecutive integers. Th sum is equal to 437. Putting this information together results in the equation:

$$x + x + 1 = 437$$

To solve, first collect like terms to obtain:

$$2x + 1 = 437$$

Then, subtract 1 from both sides and then divide by 2. The solution to the equation is $x = 218$. Therefore, the two consecutive page numbers that satisfy the problem are 218 and 219.

It is always important to make sure that answers to real-world problems make sense. For instance, it should be a red flag if the solution to this same problem resulted in decimals, which would indicate the need to check the work. Page numbers are whole numbers; therefore, if decimals are found to be answers, the solution process should be double-checked for mistakes.

Solving Equations in One Variable

An **equation in one variable** is a mathematical statement where two algebraic expressions in one variable, usually x, are set equal. To solve the equation, the variable must be isolated on one side of the equals sign. The addition and multiplication principles of equality are used to isolate the variable. The **addition principle of equality** states that the same number can be added to or subtracted from both sides of an equation. Because the same value is being used on both sides of the equals sign, equality is maintained.

For example, the equation $2x - 3 = 5x$ is equivalent to both:

$$(2x - 3) + 3 = 5x + 3$$

and

$$(2x - 3) - 5 = 5x - 5$$

This principle can be used to solve the following equation: $x + 5 = 4$. The variable x must be isolated, so to move the 5 from the left side, subtract 5 from both sides of the equals sign. Therefore:

$$x + 5 - 5 = 4 - 5$$

So, the solution is $x = -1$. This process illustrates the idea of an **additive inverse** because subtracting 5 is the same as adding -5. Basically, add the opposite of the number that must be removed to both sides of the equals sign.

The multiplication principle of equality states that equality is maintained when both sides of an equation are multiplied or divided by the same number. For example, $4x = 5$ is equivalent to both $16x = 20$ and $x = \frac{5}{4}$. Multiplying both sides times 4 and dividing both sides by 4 maintains equality. Solving the equation $6x - 18 = 5$ requires the use of both principles. First, apply the addition principle to add 18 to both sides of the equals sign, which results in $6x = 23$. Then use the multiplication principle to divide both sides by 6, giving the solution $x = \frac{23}{6}$. Using the multiplication principle in the solving process is the same as involving a multiplicative inverse. A **multiplicative inverse** is a value that, when multiplied by a given number, results in 1. Dividing by 6 is the same as multiplying by $\frac{1}{6}$, which is both the reciprocal and multiplicative inverse of 6.

When solving linear equations, check the answer by plugging the solution back into the original equation. If the result is a false statement, something was done incorrectly during the solution procedure. Checking the example above gives the following:

$$6 \times \frac{23}{6} - 18 = 23 - 18 = 5$$

Therefore, the solution is correct.

Some equations in one variable involve fractions or the use of the distributive property. In either case, the goal is to obtain only one variable term and then use the addition and multiplication principles to isolate that variable. Consider the equation $\frac{2}{3}x = 6$. To solve for x, multiply each side of the equation by the reciprocal of $\frac{2}{3}$, which is $\frac{3}{2}$. This step results in $\frac{3}{2} \times \frac{2}{3}x = \frac{3}{2} \times 6$ which simplifies into the solution $x = 9$. Now consider the equation:

$$3(x + 2) - 5x = 4x + 1$$

Use the distributive property to clear the parentheses. Therefore, multiply each term inside the parentheses by 3. This step results in:

$$3x + 6 - 5x = 4x + 1$$

Next, collect like terms on the left-hand side. Like terms are terms with the same variable or variables raised to the same exponent(s). Only like terms can be combined through addition or subtraction. After collecting like terms, the equation is:

$$-2x + 6 = 4x + 1$$

Finally, apply the addition and multiplication principles. Add $2x$ to both sides to obtain:

$$6 = 6x + 1$$

Then, subtract 1 from both sides to obtain $5 = 6x$. Finally, divide both sides by 6 to obtain the solution $\frac{5}{6} = x$.

Two other types of solutions can be obtained when solving an equation in one variable. There could be no solution, or the solution set could contain all real numbers. Consider the equation:

$$4x = 6x + 5 - 2x$$

First, the like terms can be combined on the right to obtain;

$$4x = 4x + 5$$

Next, subtract $4x$ from both sides. This step results in the false statement $0 = 5$. There is no value that can be plugged into x that will ever make this equation true. Therefore, there is no solution. The solution procedure contained correct steps, but the result of a false statement means that no value satisfies the equation. The symbolic way to denote that no solution exists is $\emptyset$.

Next, consider the equation:

$$5x + 4 + 2x = 9 + 7x - 5$$

Combining the like terms on both sides results in:

$$7x + 4 = 7x + 4$$

The left-hand side is exactly the same as the right-hand side. Using the addition principle to move terms, the result is $0 = 0$, which is always true. Therefore, the original equation is true for any number, and the solution set is all real numbers. The symbolic way to denote such a solution set is $\mathbb{R}$, or in interval notation, $(-\infty, \infty)$.

Solving a Linear Inequality in One Variable

A **linear equation** in x can be written in the form $ax + b = 0$. A **linear inequality** is very similar, although the equals sign is replaced by an inequality symbol such as $<$, $>$, $\leq$, or $\geq$. In any case, a can never be 0. Some examples of linear inequalities in one variable are:

$$2x + 3 < 0$$

and

$$4x - 2 \leq 0$$

Solving an inequality involves finding the set of numbers that, when plugged into the variable, makes the inequality a true statement.

These numbers are known as the **solution set** of the inequality. To solve an inequality, use the same properties that are necessary in solving equations. First, add or subtract variable terms and/or constants to obtain all variable terms on one side of the equals sign and all constant terms on the other side. Then, either multiply or divide both sides by the same number to obtain an inequality that gives the solution set. When multiplying or dividing by a negative number, change the direction of the inequality symbol. The solution set can be graphed on a number line.

Consider the linear inequality:

$$-2x - 5 > x + 6$$

First, add 5 to both sides and subtract x from both sides to obtain $-3x > 11$. Then, divide both sides by -3, making sure to change the direction of the inequality symbol. These steps result in the solution $x < -\frac{11}{3}$. Therefore, any number less than $-\frac{11}{3}$ satisfies this inequality.

Translating Phrases and Sentences into Expressions, Equations, and Inequalities

When presented with a real-world problem, the first step is to determine what unknown quantity must be solved for. Use a variable, such as x or t, to represent that unknown quantity. Sometimes, there can be two or more unknown quantities. In this case, either choose an additional variable, or if a relationship exists between the unknown quantities, express the other quantities in terms of the original variable. After choosing the variables, form algebraic expressions and/or equations that represent the verbal statement in the problem. The following table shows examples of vocabulary used to represent the different operations:

Addition	Sum, plus, total, increase, more than, combined, in all
Subtraction	Difference, less than, subtract, reduce, decrease, fewer, remain
Multiplication	Product, multiply, times, part of, twice, triple
Division	Quotient, divide, split, each, equal parts, per, average, shared

The combination of operations and variables form both mathematical expression and equations. As mentioned, the difference between expressions and equations are that there is no equals sign in an expression, and that expressions are evaluated to find an unknown quantity, while equations are solved to find an unknown quantity. Also, inequalities can exist within verbal mathematical statements. Instead of a statement of equality, expressions state quantities are *less than, less than or equal to, greater than,* or *greater than or equal to*. Another type of inequality is when a quantity is said to be not equal to another quantity ($\neq$).

The steps for solving inequalities in one variable are the same steps for solving equations in one variable. The addition and multiplication principles are used. However, to maintain a true statement when using the $<$, $\leq$, $>$, and $\geq$ symbols, if a negative number is either multiplied times both sides of an inequality or divided from both sides of an inequality, the sign must be flipped. For instance, consider the following inequality: $3 - 5x \leq 8$. First, 3 is subtracted from each side to obtain $-5x \leq 5$. Then, both sides are divided by -5, while flipping the sign, to obtain $x \geq -1$. Therefore, any real number greater than or equal to -1 satisfies the original inequality.

Rewriting Simple Rational Expressions

A **rational expression** is a fraction or a ratio in which both the numerator and denominator are polynomials that are not equal to zero. A polynomial is a mathematical expression containing addition, subtraction, or multiplication of one or more constants multiplied by variables raised to positive powers. Here are some examples of rational expressions: $\frac{2x^2+6x}{x}$, $\frac{x-2}{x^2-6x+8}$, and $\frac{x^3-1}{x+2}$. Such expressions can be simplified using different forms of division. The first example can be simplified in two ways. Then, cancelling out an x in each numerator and the x in each denominator results in $2x + 6$. It also can be simplified using factoring and then crossing out common factors in the numerator and denominator. For instance, it can be written as:

$$\frac{2x(x+3)}{x} = 2(x+3) = 2x + 6$$

The second expression above can also be simplified using factoring. It can be written as:

$$\frac{x-2}{(x-2)(x-4)} = \frac{1}{x-4}$$

Finally, the third example can only be simplified using long division, as there are no common factors in the numerator and denominator. First, divide the first term of the numerator by the first term of the denominator, then write the result in the quotient. Then, multiply the divisor times that number and write it below the dividend. Subtract and continue the process until each term in the divisor is accounted for. Here is the actual long division:

Simplifying Expressions Using Long Division

$$
\begin{array}{r}
x^2 \quad -2x \quad +4 \\
x+2 \overline{\smash{\big)}\, x^3 \qquad\qquad\quad -1} \\
\underline{x^3 \quad +2x^2} \\
-2x^2 \qquad\qquad -1 \\
\underline{-2x^2 \quad -4x} \\
4x \quad -1 \\
\underline{4x \quad +8} \\
-9
\end{array}
$$

Function Notation

A **relation** is any set of ordered pairs (x, y). The values listed first in the ordered pairs, known as the x-coordinates, make up the domain of the relation. The values listed second, known as the y-coordinates, make up the range. A relation in which every member of the domain corresponds to only one member of the range is known as a **function**. A function cannot have a member of the domain corresponding to two members of the range. Functions are most often given in terms of equations instead of ordered pairs. For instance, here is an equation of a line: $y = 2x + 4$. In function notation, this can be written as

$$f(x) = 2x + 4$$

The expression $f(x)$ is read "f of x" and it shows that the inputs, the x-values, get plugged into the function and the output is $y = f(x)$. The set of all inputs are in the domain, and the set of all outputs are in the range.

The x-values are known as the **independent variables** of the function and the y-values are known as the **dependent variables** of the function. The y-values depend on the x-values. For instance, if $x = 2$ is plugged into the function shown above, the y-value depends on that input.

$$f(2) = 2 \times 2 + 4 = 8.$$

Therefore, $f(2) = 8$, which is the same as writing the ordered pair $(2, 8)$. To graph a function, graph it in equation form and plot ordered pairs.

114

Due to the definition of a function, the graph of a function cannot have two of the same x-components paired to different y-component. For example, the ordered pairs $(3, 4)$ and $(3, -1)$ cannot be in a valid function. Therefore, all graphs of functions pass the **vertical line test**. If any vertical line intersects a graph in more than one place, the graph is not that of a function. For instance, the graph of a circle is not a function because one can draw a vertical line through a circle and intersect the circle twice. Common functions include lines and polynomials, which pass the vertical line test.

Linear Functions that Model a Linear Relationship

A **linear function that models a linear relationship between two quantities** is of the form $y = mx + b$, or in function form $f(x) = mx + b$. In a linear function, the value of y depends on the value of x, and y increases or decreases at a constant rate as x increases. Therefore, the independent variable is x, and the dependent variable is y. The graph of a linear function is a line, and the constant rate can be seen by looking at the steepness, or slope, of the line. If the line increases from left to right, the slope is positive. If the line slopes downward from left to right, the slope is negative. In the function, m represents slope. Each point on the line is an **ordered pair** (x, y), where x represents the x-coordinate of the point and y represents the y-coordinate of the point. The point where $x = 0$ is known as the y-intercept, and it is the place where the line crosses the y-axis. If $x = 0$ is plugged into $f(x) = mx + b$, the result is $f(0) = b$, so therefore, the point $(0, b)$ is the y-intercept of the line. The derivative of a linear function is its slope.

Consider the following situation. A taxicab driver charges a flat fee of $2 per ride and $3 a mile. This statement can be modeled by the function $f(x) = 3x + 2$ where x represents the number of miles and $f(x) = y$ represents the total cost of the ride. The total cost increases at a constant rate of $2 per mile, and that is why this situation is a linear relationship. The slope $m = 3$ is equivalent to this rate of change. The flat fee of $2 is the y-intercept. It is the place where the graph crosses the x-axis, and it represents the cost when $x = 0$, or when no miles have been traveled in the cab. The y-intercept in this situation represents the flat fee.

Polynomial Functions

A **polynomial function** is a function containing a polynomial expression, which is an expression containing constants and variables combined using the four mathematical operations. The degree of a polynomial depends on the largest exponent in the expression. Typical polynomial functions are **quartic**, with a degree of 4, **cubic**, with a degree of 3, and **quadratic**, with a degree of 2. Note that the exponents on the variables can only be nonnegative integers. The domain of any polynomial function is all real numbers because any number plugged into a polynomial expression grants a real number output.

An example of a quartic polynomial equation is:

$$y = x^4 + 3x^3 - 2x + 1$$

The zeros of a polynomial function are the points where its graph crosses the y-axis. In order to find the number of real zeros of a polynomial function, use **Descartes' Rule of Signs**, which states that the number of possible positive real zeros is equal to the number of sign changes in the coefficients. If there is only one sign change, there is only one positive real zero. In the example above, the signs of the coefficients are positive, positive, negative, and positive. The sign changes twice; therefore, there are at most two positive real zeros. The number of possible negative real zeros is equal to the number of sign changes in the coefficients when plugging $-x$ into the equation. Again, if there is only one sign change, there is only one negative real zero. The polynomial result when plugging $-x$ into the equation is:

$$y = (-x)^4 + 3(-x)^3 - 2(-x) + 1$$

$$y = x^4 - 3x^3 + 2x + 1$$

The sign changes two times, so there are at most two negative real zeros. Another polynomial equation this rule can be applied to is:

$$y = x^3 + 2x - x - 5$$

There is only one sign change in the terms of the polynomial, so there is exactly one real zero. When plugging $-x$ into the equation, the polynomial result is:

$$-x^3 - 2x - x - 5$$

There are no sign changes in this polynomial, so there are no possible negative zeros.

Adding, Subtracting, and Multiplying Polynomial Equations

When working with polynomials, **like terms** are terms that contain the same variables with the same powers. For example, x^4y^5 and $9x^4y^5$ are like terms. The coefficients are different, but the same variables are raised to the same powers. When adding polynomials, only terms that are like can be added. When adding two like terms, just add the coefficients and leave the variables alone. This process uses the distributive property. For example:

$$x^4y^5 + 9x^4y^5 = (1 + 9)x^4y^5 = 10x^4y^5$$

Therefore, when adding two polynomials, simply add the like terms together. Unlike terms cannot be combined.

Subtracting polynomials involves adding the opposite of the polynomial being subtracted. Basically, the sign of each term in the polynomial being subtracted is changed, and then the like terms are combined because it is now an addition problem. For example, consider the following:

$$6x^2 - 4x + 2 - (4x^2 - 8x + 1)$$

Add the opposite of the second polynomial to obtain:

$$6x^2 - 4x + 2 + (-4x^2 + 8x - 1)$$

Then, collect like terms to obtain:

$$2x^2 + 4x + 1$$

Multiplying polynomials involves using the product rule for exponents that:

$$b^m b^n = b^{m+n}$$

Basically, when multiplying expressions with the same base, just add the exponents. Multiplying a monomial by a monomial involves multiplying the coefficients together and then multiplying the variables together using the product rule for exponents.

For instance,

$$8x^2 y \times 4x^4 y^2 = 32x^6 y^3$$

When multiplying a monomial by a polynomial that is not a monomial, use the distributive property to multiply each term of the polynomial times the monomial.

For example,

$$3x(x^2 + 3x - 4) = 3x^3 + 9x^2 - 12x$$

Finally, multiplying two polynomials when neither one is a monomial involves multiplying each term of the first polynomial times each term of the second polynomial. There are some shortcuts, given certain scenarios. For instance, a binomial times a binomial can be found by using the **FOIL (Firsts, Outers, Inners, Lasts)** method shown here:

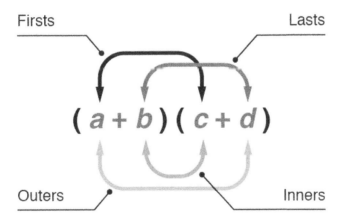

Finding the product of a sum and difference of the same two terms is simple because if it was to be foiled out, the outer and inner terms would cancel out. For instance:

$$(x + y)(x - y) = x^2 + xy - xy - y^2$$

Finally, the square of a binomial can be found using the following formula:

$$(a \pm b)^2 = a^2 \pm 2ab + b^2$$

Relationship Between Zeros and Factors of Polynomials

A polynomial is a mathematical expression containing addition, subtraction, or multiplication of one or more constants multiplied by variables raised to positive powers. A **polynomial equation** is a polynomial set equal to another polynomial, or in standard form, a polynomial is set equal to zero. A **polynomial function** is a polynomial set equal to y. For instance, $x^2 + 2x - 8$ is a polynomial, $x^2 + 2x - 8 = 0$ is a polynomial equation, and $y = x^2 + 2x - 8$ is the corresponding polynomial function. To solve a polynomial equation, the x-values in which the graph of the corresponding polynomial function crosses the x-axis are sought.

These coordinates are known as the **zeros** of the polynomial function because they are the coordinates in which the y-coordinates are 0. One way to find the zeros of a polynomial is to find its factors, then set each individual factor equal to 0, and solve each equation to find the zeros. A **factor** is a linear expression, and to completely factor a polynomial, the polynomial must be rewritten as a product of individual linear factors. The polynomial listed above can be factored as $(x + 4)(x - 2)$. Setting each factor equal to zero results in the zeros $x = -4$ and $x = 2$.

Here is the graph of the zeros of the polynomial:

The Graph of the Zeros of $x^2 + 2x - 8 = 0$

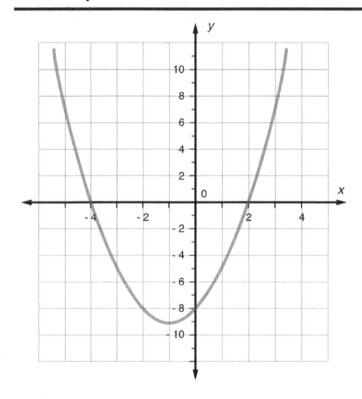

Interpreting Solutions of Multistep One-Variable Linear Equations and Inequalities

Multistep one-variable equations involve the use of one variable in an equation with many operations. For example, the equation $2x + 4 = 10$ involves one variable, x, and multiple steps to solve for the value of x. The first step is to move the four to the opposite side of the equation by subtracting 4. The next step is to divide by 2. The final answer yields a value of 3 for the variable x. The steps for this process are shown below:

$$2x + 4 = 10$$

Subtract 4 on both sides

$$2x = 6$$

Divide by 2 on both sides

$$x = 3$$

When the result is found, the value of the variable must be interpreted. For this problem, a value of 3 can be understood as the number that can be doubled and then increased by 4 to yield a value of 10.

Inequalities can also be interpreted in much the same way. The following inequality can be solved to find the value of b:

$$\frac{b}{7} - 8 \geq 7$$

This inequality models the amount of money a group of friends earned for cleaning up a neighbor's yard, b. There were 7 friends, so the money had to be split seven times. Then $8 was taken away from each friend to pay for materials they bought to help clean the yard. All these things needed to be less than or equal to seven for the friends to each receive at least $7. The first step is to add 8 to both sides of the inequality. Then, both sides can be multiplied by 7 to get rid of the denominator on the left side. The resulting inequality is $b \geq 105$. Because the answer is not only an equals sign, the value for b is not a single number.

In this problem, the answer communicates that the value of b must be greater than or equal to $105 in order for each friend to make at least $7 for their work. The number for b, what they are paid, can be more than 105 because that would mean they earned more money. They do not want it to be less than 105 because their profit will drop below $7 per piece.

Linear Relationships Represented by Graphs, Equations, and Tables

Graphs, equations, and tables are three different ways to represent linear relationships. The following graph shows a linear relationship because the relationship between the two variables is constant. Each time the distance increases by 25 miles, 1 hour passes. This pattern continues for the rest of the graph. The line represents a constant rate of 25 miles per hour. This graph can also be used to solve problems involving predictions for a future time. After 8 hours of travel, the rate can be used to predict the distance covered. Eight hours of travel at 25 miles per hour covers a distance of 200 miles. The equation at the top of the graph corresponds to this rate also. The same prediction of distance in a given time can be found using the equation.

For a time of 10 hours, the distance would be 250 miles, as the equation yields:

$$d = 25 \times 10 = 250$$

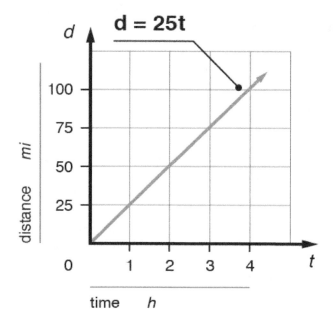

Another representation of a linear relationship can be seen in a table. In the table below, the y-values increase by 3 as the x-values increase by 1. This pattern shows that the relationship is linear. If this table shows the money earned, y-value, for the hours worked, x-value, then it can be used to predict how much money will be earned for future hours. If 6 hours are worked, then the pay would be $19. For further hours and money to be determined, it would be helpful to have an equation that models this table of values. The equation will show the relationship between x and y. The y-value can be determined by multiplying the x-value by 3, then adding 1. The following equation models this relationship:

$$y = 3x + 1$$

Now that there is an equation, any number of hours, x, can be substituted into the equation to find the amount of money earned, y.

y = 3x + 1	
x	y
0	1
1	4
2	7
4	13
5	16

Graphing and Statistics

Coordinate Plane

The coordinate plane is a way of identifying the position of a point in relation to two axes. The **coordinate plane** is made up of two intersecting lines, the x-axis and the y-axis. These lines intersect at a right angle, and their intersection point is called the **origin**. The points on the coordinate plane are labeled based on their position in relation to the origin. If a point is found 4 units to the right and 2 units up from the origin, the location is described as $(4, 2)$. These numbers are the x- and y-coordinates, always written in the order (x, y). This point is also described as lying in the first quadrant. Every point in the first quadrant has a location that is positive in the x and y directions.

The following figure shows the coordinate plane with examples of points that lie in each quadrant.

The Coordinate Plane

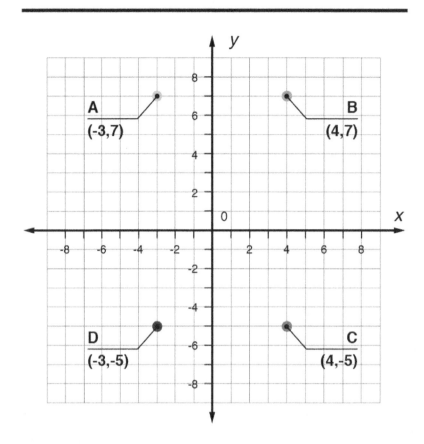

Point B lies in the first quadrant, described positive x- and y-values, above the x-axis and to the right of the y-axis. Point A lies in the second quadrant, where the x-value is negative and y-value is positive. This quadrant is above the x-axis and to the left of the y-axis. Point D lies in the third quadrant, where both the x- and y-values are negative. This quadrant is below the x-axis and to the left of the y-axis. Point C is in the fourth quadrant, where the x-value is positive and the y-value is negative.

Tables, Charts, and Graphs

They all organize, categorize, and compare data, and they come in different shapes and sizes. Each type has its own way of showing information, whether through a column, shape, or picture. To answer a question relating to a table, chart, or graph, some steps should be followed. First, the problem should be read thoroughly to determine what is being asked to determine what quantity is unknown. Then, the title of the table, chart, or graph should be read. The title should clarify what data is actually being summarized in the table. Next, look at the key and labels for both the horizontal and vertical axes, if they are given. These items will provide information about how the data is organized. Finally, look to see if there is any more labeling inside the table. Taking the time to get a good idea of what the table is summarizing will be helpful as it is used to interpret information.

Tables are a good way of showing a lot of information in a small space. The information in a table is organized in columns and rows. For example, a table may be used to show the number of votes each candidate received in an election. By interpreting the table, one may observe which candidate won the election and which candidates came in second and third. In using a bar chart to display monthly rainfall amounts in different countries, rainfall can be compared between countries at different times of the year. Graphs are also a useful way to show change in variables over time, as in a line graph, or percentages of a whole, as in a pie graph.

The table below relates the number of items to the total cost. The table shows that one item costs $5. By looking at the table further, five items cost $25, ten items cost $50, and fifty items cost $250. This cost can be extended for any number of items. Since one item costs $5, then two items would cost $10. Though this information is not in the table, the given price can be used to calculate unknown information.

Number of Items	1	5	10	50
Cost ($)	5	25	50	250

A **bar graph** is a graph that summarizes data using bars of different heights. It is useful when comparing two or more items or when seeing how a quantity changes over time. It has both a horizontal and vertical axis. To interpret bar graphs, recognize what each bar represents and connect that to the two variables. The bar graph below shows the scores for six people during three different games. The different colors of the bars distinguish between the three games, and the height of the bar indicates their score for that game. William scored 25 on game 3, and Abigail scored 38 on game 3. By comparing the bars, it is obvious that Williams scored lower than Abigail.

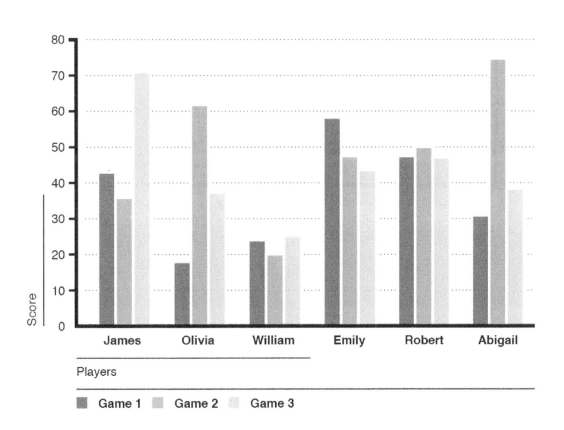

123

A line graph is a way to compare two variables that are plotted on opposite axes of a graph. The line indicates a continuous change as it rises or falls. The line's rate of change is known as its slope. The horizontal axis often represents a variable of time. Audiences can quickly see if an amount has increased or decreased over time. The bottom of the graph, or the *x*-axis, shows the units for time, such as days, hours, months, etc.

If there are multiple lines, a comparison can be made between what the two lines represent. For example, the following line graph shows the change in temperature over five days. The top line represents the high, and the bottom line represents the low for each day. Looking at the top line alone, the high decreases for a day, then increases on Wednesday. Then it decreases on Thursday and increases again on Friday. The low temperatures have a similar trend, shown in the bottom line.

The range in temperatures each day can also be calculated by finding the difference between the top line and bottom line on a particular day. On Wednesday, the range was 14 degrees, from 62 to 76° F.

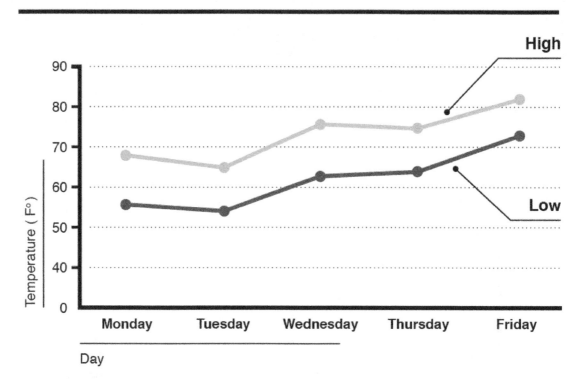

Daily Temperatures

Pie charts are used to show percentages of a whole, as each category is given a piece of the pie, and together all the pieces make up a whole. They are a circular representation of data which are used to highlight numerical proportion. It is true that the arc length of each pie slice is proportional to the amount it individually represents. When a pie chart is shown, an audience can quickly make comparisons by comparing the sizes of the pieces of the pie. They can be useful for comparison between different categories. The following pie chart is a simple example of three different categories shown in comparison to each other.

Light gray represents cats, dark gray represents dogs, and the medium shade of gray represents other pets. These three equal pieces each represent just more than 33 percent, or $\frac{1}{3}$ of the whole. Values 1 and 2 may be combined to represent $\frac{2}{3}$ of the whole.

In an example where the total pie represents 75,000 animals, then cats would be equal to $\frac{1}{3}$ of the total, or 25,000. Dogs would equal 25,000 and other pets also equal 25,000.

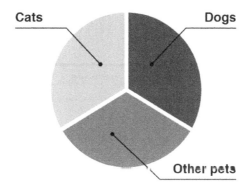

The fact that a circle is 360 degrees is used to create a pie chart. Because each piece of the pie is a percentage of a whole, that percentage is multiplied times 360 to get the number of degrees each piece represents. In the example above, each piece is $\frac{1}{3}$ of the whole, so each piece is equivalent to 120 degrees. Together, all three pieces add up to 360 degrees.

Stacked bar graphs, also used fairly frequently, are used when comparing multiple variables at one time. They combine some elements of both pie charts and bar graphs, using the organization of bar graphs and the proportionality aspect of pie charts. The following is an example of a stacked bar graph that represents the number of students in a band playing drums, flute, trombone, and clarinet. Each bar graph is broken up further into girls and boys:

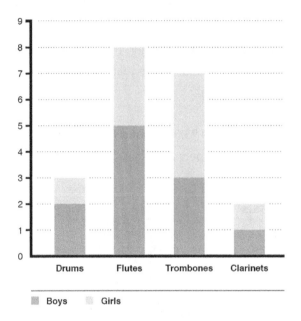

To determine how many boys play trombone, refer to the darker portion of the trombone bar, resulting in 3 students.

A **scatterplot** is another way to represent paired data. It uses Cartesian coordinates, like a line graph, meaning it has both a horizontal and vertical axis. Each data point is represented as a dot on the graph. The dots are never connected with a line. For example, the following is a scatterplot showing people's height versus age.

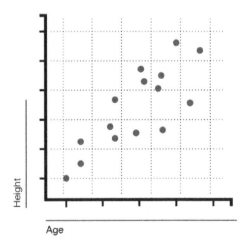

Mathematics Knowledge

A scatterplot, also known as a **scattergram**, can be used to predict another value and to see if an association, known as a **correlation**, exists between a set of data. If the data resembles a straight line, the data is associated. The following is an example of a scatterplot in which the data does not seem to have an association:

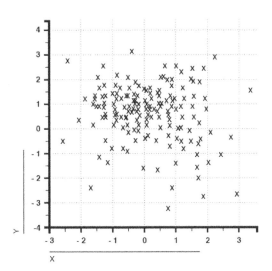

Sets of numbers and other similarly organized data can also be represented graphically. Venn diagrams are a common way to do so. A **Venn diagram** represents each set of data as a circle. The circles overlap, showing that each set of data is overlapping. A Venn diagram is also known as a **logic diagram** because it visualizes all possible logical combinations between two sets. Common elements of two sets are represented by the area of overlap.

The following is an example of a Venn diagram of two sets A and B:

Parts of the Venn Diagram

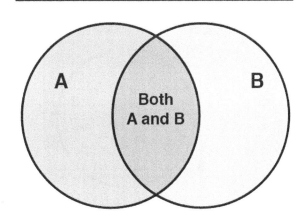

Another name for the area of overlap is the **intersection**. The intersection of A and B, $A \cap B$, contains all elements that are in both sets A and B. The **union** of A and B, $A \cup B$, contains all elements in both sets A and B. Finally, the **complement** of $A \cup B$ is equal to all elements that are not in either set A or set B. These elements are placed outside of the circles.

The following is an example of a Venn diagram in which 30 students were surveyed asking which type of siblings they had: brothers, sisters, or both. Ten students only had a brother, seven students only had a sister, and five had both a brother and a sister. Therefore, five is the intersection, represented by the section where the circles overlap. Two students did not have a brother or a sister. Two is therefore the complement and is placed outside of the circles.

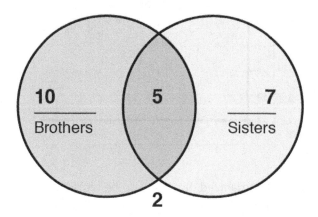

Venn diagrams can have more than two sets of data. The more circles, the more logical combinations are represented by the overlapping. The following is a Venn diagram that represents a different situation. Now, there were 30 students surveyed about the color of their socks. The innermost region represents those students that have green, pink, and blue socks on (perhaps a striped pattern). Therefore, two students had all three colors on their socks. In this example, all students had at least one of the three colors on their socks, so no one exists in the complement.

30 students

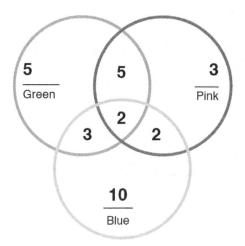

Venn diagrams are typically not drawn to scale; however, if they are, and if each circle's area is proportional to the amount of data it represents, then it is called an area-proportional Venn diagram.

Evaluating the Information in Tables, Charts, and Graphs Using Statistics

One way information can be interpreted from tables, charts, and graphs is through statistics. The three most common calculations for a set of data are the mean, median, and mode. These three are called **measures of central tendency**, which are helpful in comparing two or more different sets of data. The **mean** refers to the average and is found by adding up all values and dividing the total by the number of values. In other words, the mean is equal to the sum of all values divided by the number of data entries. For example, if you bowled a total of 532 points in 4 bowling games, your mean score was $\frac{532}{4} = 133$ points per game. Students can apply the concept of mean to calculate what score they need on a final exam to earn a desired grade in a class.

The **median** is found by lining up values from least to greatest and choosing the middle value. If there is an even number of values, then calculate the mean of the two middle amounts to find the median. For example, the median of the set of dollar amounts $5, $6, $9, $12, and $13 is $9. The median of the set of dollar amounts $1, $5, $6, $8, $9, $10 is $7, which is the mean of $6 and $8. The **mode** is the value that occurs the most. The mode of the data set {1, 3, 1, 5, 5, 8, 10} actually refers to two numbers: 1 and 5. In this case, the data set is bimodal because it has two modes. A data set can have no mode if no amount is repeated. Another useful statistic is range. The **range** for a set of data refers to the difference between the highest and lowest value.

In some cases, numbers in a list of data might have weights attached to them. In that case, a weighted mean can be calculated. A common application of a weighted mean is GPA. In a semester, each class is assigned a number of credit hours, its weight, and at the end of the semester each student receives a grade. To compute GPA, an A is a 4, a B is a 3, a C is a 2, a D is a 1, and an F is a 0. Consider a student that takes a 4-hour English class, a 3-hour math class, and a 4-hour history class and receives all B's. The weighted mean, GPA, is found by multiplying each grade times its weight, number of credit hours, and dividing by the total number of credit hours. Therefore, the student's GPA is:

$$\frac{3 \times 4 + 3 \times 3 + 3 \times 4}{11} = \frac{33}{11} = 3.0.$$

The following bar chart shows how many students attend a cycle class on each day of the week. To find the mean attendance for the week, add each day's attendance together:

$$10 + 7 + 6 + 9 + 8 + 14 + 4 = 58$$

Then divide the total by the number of days, $58 \div 7 = 8.3$. The mean attendance for the week was 8.3 people. The median attendance can be found by putting the attendance numbers in order from least to greatest: 4, 6, 7, 8, 9, 10, 14, and choosing the middle number: 8 people. There is no mode for this set of data because no numbers repeat. The range is 10, which is found by finding the difference between the lowest number, 4, and the highest number, 14.

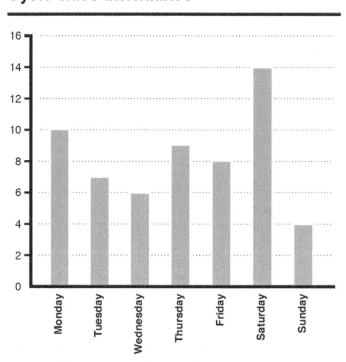

Cycle class attendance

A **histogram** is a bar graph used to group data into "bins" that cover a range on the horizontal, or x-axis. Histograms consist of rectangles whose heights are equal to the frequency of a specific category. The horizontal axis represents the specific categories. Because they cover a range of data, these bins have no gaps between bars, unlike the bar graph above. In a histogram showing the heights of adult golden retrievers, the bottom axis would be groups of heights, and the y-axis would be the number of dogs in each range. Evaluating this histogram would show the height of most golden retrievers as falling within a certain range. It also provides information to find the average height and range for how tall golden retrievers may grow.

The following is a histogram that represents exam grades in a given class. The horizontal axis represents ranges of the number of points scored, and the vertical axis represents the number of students. For example, approximately 33 students scored in the 60 to 70 range.

Results of the exam

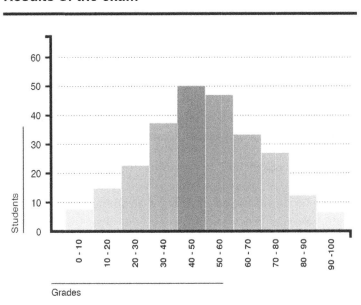

Certain measures of central tendency can be easily visualized with a histogram. If the points scored were shown with individual rectangles, the tallest rectangle would represent the mode. A **bimodal** set of data would have two peaks of equal height. Histograms can be classified as having data **skewed to the left**, **skewed to the right**, or **normally distributed**, which is also known as **bell-shaped**.

These three classifications can be seen in the following image:

Measures of central tendency images

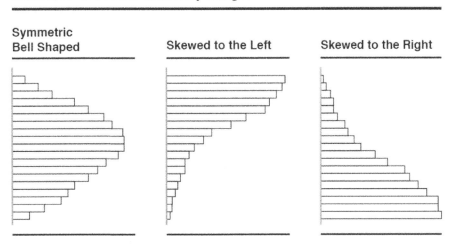

131

When the data follows the normal distribution, the mean, median, and mode are all very close. They all represent the most typical value in the data set. In this case, the mean is typically considered the best measure of central tendency because it includes all data points. However, if the data is skewed, the mean becomes less meaningful because it is dragged in the direction of the skew. Therefore, the median becomes the best measure because it is not affected by any outliers.

The measures of central tendency and the range may also be found by evaluating information on a line graph.

In the line graph from a previous example that showed the daily high and low temperatures, the average high temperature can be found by gathering data from each day on the triangle line. The days' highs are 69, 65, 75, 74, and 81. To find the average, add them together to get 364, then divide by 5 (because there are 5 temperatures). The average high for the five days is 72.8. If 72.8 degrees is found on the graph, it will fall in the middle of all the values. The average low temperature can be found in the same way.

Given a set of data, the **correlation coefficient**, r, measures the association between all the data points. If two values are correlated, there is an association between them. However, correlation does not necessarily mean causation, or that one value causes the other. There is a common mistake made that assumes correlation implies causation. Average daily temperature and number of sunbathers are both correlated and have causation. If the temperature increases, that change in weather causes more people to want to catch some rays. However, wearing plus-size clothing and having heart disease are two variables that are correlated but do not have causation. The larger someone is, the more likely he or she is to have heart disease. However, being overweight does not cause someone to have the disease.

The value of the correlation coefficient is between -1 and 1, where -1 represents a perfect negative linear relationship, 0 represents no relationship between the two data sets, and 1 represents a perfect positive linear relationship. A negative linear relationship means that as x-values increase, y-values decrease. A positive linear relationship means that as x-values increase, y-values increase. The formula for computing the correlation coefficient is:

$$r = \frac{n \sum xy - (\sum x)(\sum y)}{\sqrt{n(\sum x^2) - (\sum x)^2}\sqrt{n(\sum y^2) - (y)^2}}$$

In this formula, n is the number of data points.

The closer r is to 1 or -1, the stronger the correlation. A correlation can be seen when plotting data. If the graph resembles a straight line, there is a correlation.

Constructing Graphs That Correctly Represent Given Data

Data is often displayed with a line graph, bar graph, or pie chart.

The line graph below shows the number of push-ups that a student did over one week:

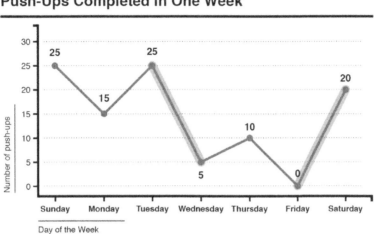

Push-Ups Completed in One Week

Notice that the horizontal axis displays the day of the week and the vertical axis displays the number of push-ups. A point is placed above each day of the week to show how many push-ups were done each day. For example, on Sunday the student did 25 push-ups. The line that connects the points shows how much the number of push-ups fluctuated throughout the week.

The bar graph below compares the number of people who own various types of pets:

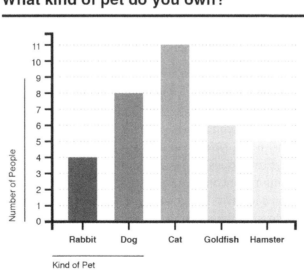

What kind of pet do you own?

On the horizontal axis, the kind of pet is displayed. On the vertical axis, the number of people is displayed. Bars are drawn to show the number of people who own each type of pet. With the bar graph, it can quickly be determined that the fewest number of people own a rabbit and the greatest number of people own a cat.

The pie graph below displays students in a class who scored A, B, C, or D. Each slice of the pie is drawn to show the portion of the whole class that is represented by each letter grade. For example, the smallest portion represents students who scored a D. This means that the fewest number of students scored a D.

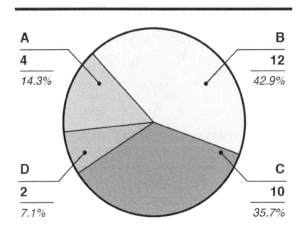

Student Grades

A
4
14.3%

B
12
42.9%

D
2
7.1%

C
10
35.7%

Choosing Appropriate Graphs to Display Data

Data may be displayed with a line graph, bar graph, or pie chart.

- A line graph is used to display data that changes continuously over time.

- A bar graph is used to compare data from different categories or groups and is helpful for recognizing relationships.

- A pie chart is used when the data represents parts of a whole.

Important Features of Graphs

A **graph** is a pictorial representation of the relationship between two variables. To read and interpret a graph, it is necessary to identify important features of the graph. First, read the title to determine what data sets are being related in the graph. Next, read the axis labels and understand the scale that is used. The horizontal axis often displays categories, like years, months, or types of pets. The vertical axis often displays numerical data like amount of income, number of items sold, or number of pets owned. Check to see what increments are used on each axis. The changes on the axis may represent fives, tens, hundreds, or any increment. Be sure to note what the increment is because it will affect the interpretation of the graph. Now, locate on the graph an element of interest and move across to find the element to which it relates. For example, notice an element displayed on the horizontal axis, find that element on the graph, and then follow it across to the corresponding point on the vertical axis. Using the appropriate scale, interpret the relationship.

Explaining the Relationship between Two Variables

Independent and dependent are two types of variables that describe how they relate to each other. The **independent variable** is the variable controlled by the experimenter. It stands alone and is not changed by other parts of the experiment. This variable is normally represented by x and is found on the horizontal, or x-axis, of a graph. The **dependent variable** changes in response to the independent variable. It reacts to, or depends on, the independent variable. This variable is normally represented by y and is found on the vertical, or y-axis of the graph.

The relationship between two variables, x and y, can be seen on a scatterplot.

The following scatterplot shows the relationship between weight and height. The graph shows the weight as x and the height as y. The first dot on the left represents a person who is 45 kg and approximately 150 cm tall. The other dots correspond in the same way. As the dots move to the right and weight increases, height also increases. A line could be drawn through the middle of the dots to move from bottom left to top right. This line would indicate a **positive correlation** between the variables. If the variables had a **negative correlation**, then the dots would move from the top left to the bottom right.

Height and Weight

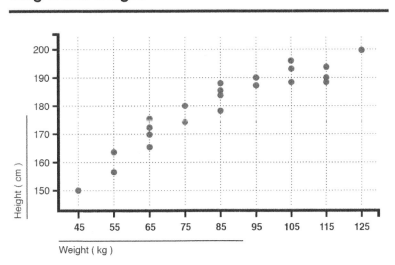

A scatterplot is useful in determining the relationship between two variables, but it is not required. Consider an example where a student scores a different grade on his math test for each week of the month. The independent variable would be the weeks of the month. The dependent variable would be the grades because they change depending on the week. If the grades trended up as the weeks passed, then the relationship between grades and time would be positive. If the grades decreased as the time passed, then the relationship would be negative. (As the number of weeks went up, the grades went down.)

The relationship between two variables can further be described as strong or weak. The relationship between age and height shows a strong positive correlation because children grow taller as they grow up. In adulthood, the relationship between age and height becomes weak, and the dots will spread out. People stop growing in adulthood, and their final heights vary depending on factors like genetics and health. The closer the dots on the graph, the stronger the relationship. As they spread apart, the relationship becomes weaker. If they are too spread out to determine a correlation up or down, then the variables are said to have no correlation.

Variables are values that change, so determining the relationship between them requires an evaluation of who changes them. If the variable changes because of a result in the experiment, then it's dependent. If the variable changes before the experiment, or is changed by the person controlling the experiment, then it's the independent variable. As they interact, one is manipulated by the other. The manipulator is the independent, and the manipulated is the dependent. Once the independent and dependent variable are determined, they can be evaluated to have a positive, negative, or no correlation.

Comparing Linear Growth with Exponential Growth

Linear growth involves a quantity, the dependent variable, increasing or decreasing at a constant rate as another quantity, the independent variable, increases as well. The graph of linear growth is a straight line. Linear growth is represented as the following equation: $y = mx + b$, where m is the **slope** of the line, also known as the **rate of change**, and b is the y-intercept. If the y-intercept is 0, then the linear growth is actually known as direct variation. If the slope is positive, the dependent variable increases as the independent variable increases, and if the slope is negative, the dependent variable decreases as the independent variable increases.

Exponential growth involves the dependent variable changing by a common ratio every unit increase. The equation of exponential growth is $y = a^x$ for $a > 0$, $a \neq 1$. The value a is known as the **base**. Consider the exponential equation $y = 2^x$. When x equals 1, y equals 2, and when x equals 2, y equals 4. For every unit increase in x, the value of the output variable doubles. Here is the graph of $y = 2^x$. Notice that as the dependent variable, y, gets very large, x increases slightly. This characteristic of this graph is why sometimes a quantity is said to be blowing up exponentially.

$$y = 2^x$$

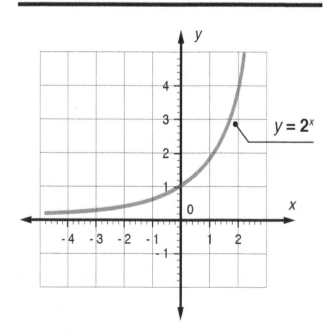

Statistical Questions

Statistics is the branch of mathematics that deals with the collection, organization, and analysis of data. A statistical question is one that can be answered by collecting and analyzing data. When collecting data, expect variability. For example, "How many pets does Yanni own?" is not a statistical question because it can be answered in one way. "How many pets do the people in a certain neighborhood own?" is a statistical question because, to determine this answer, one would need to collect data from each person in the neighborhood, and it is reasonable to expect the answers to vary.

Identify these as statistical or not statistical:

1. How old are you?

2. What is the average age of the people in your class?

3. How tall are the students in Mrs. Jones' sixth grade class?

4. Do you like Brussels sprouts?

Questions 2 and 3 are statistical questions.

Using Statistics to Investigate Data

In statistics, measures of central tendency are measures of average. They include the mean, median, mode, and midrange of a data set. The **mean**, otherwise known as the **arithmetic average**, is found by dividing the sum of all data entries by the total number of data points. The **median** is the midpoint of the data points. If there is an odd number of data points, the median is the entry in the middle. If there is an even number of data points, the median is the mean of the two entries in the middle. The **mode** is the data point that occurs most often. Finally, the **midrange** is the mean of the lowest and highest data points. Given the spread of the data, each type of measure has pros and cons.

In a **right-skewed distribution**, the bulk of the data falls to the left of the mean. In this situation, the mean is on the right of the median and the mode is on the left of the median. In a **normal distribution**, where the data are evenly distributed on both sides of the mean, the mean, median, and mode are very close to one another. In a **left-skewed distribution**, the bulk of the data falls to the right of the mean. The mean is on the left of the median and the mode is on the right of the median.

Here is an example of each type of distribution:

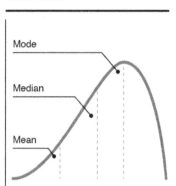
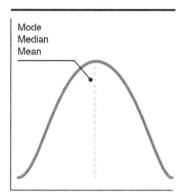
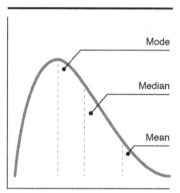

Solving Problems Involving Measures of Center and Range

A data set can be described by calculating the mean, median, and mode. These values, called **measures of center**, allow the data to be described with a single value that is representative of the data set.

The most common measure of center is the **mean**, also referred to as the **average**.

To calculate the mean,

- Add all data values together

- Divide by the sample size (the number of data points in the set)

The **median** is middle data value, so that half of the data lies below this value and half lies above it.

To calculate the median,

- Order the data from least to greatest

- The point in the middle of the set is the median

  - In the event that there is an even number of data points, add the two middle points and divide by 2

The **mode** is the data value that occurs most often.

To calculate the mode,

- Order the data from least to greatest
- Find the value that occurs most often

Example: Amelia is a leading scorer on the school's basketball team. The following data set represents the number of points that Amelia has scored in each game this season. Use the mean, median, and mode to describe the data.

16, 12, 26, 14, 28, 14, 12, 15, 25

Solution:

Mean: $16 + 12 + 26 + 14 + 28 + 14 + 12 + 15 + 25 = 162$

$162 \div 9 = 18$

Amelia averages 18 points per game.

Median: 12, 12, 14, 14, **15**, 16, 25, 26, 28

Amelia's median score is 15.

Mode: 12, 12, 14, 14, 15, 16, 25, 26, 28

12 and 14 each occur twice in the data set, so this set has 2 modes: 12 and 14.

The **range** is the difference between the largest and smallest values in the set. In the example above, the range is $28 - 12 = 16$.

ow Changes in Data Affect Measures of Center or Range

An **outlier** is a data point that lies an unusual distance from other points in the data set. Removing an outlier from a data set will change the measures of central tendency. Removing a large outlier from a data set will decrease both the mean and the median. Removing a small outlier from a data set will increase both the mean and the median. For example, given the data set {3, 6, 8, 12, 13, 14, 60}, the data point 60 is an outlier because it is unusually far from the other points. In this data set, the mean is 16.6. Notice that this mean number is even larger than all other data points in the set except for 60. Removing the outlier changes the mean to 9.3, and the median goes from 12 to 10. Removing an outlier will also decrease the range. In the data set above, the range is 57 when the outlier is included, but it decreases to 11 when the outlier is removed.

Adding an outlier to a data set will also affect the measures of central tendency. When a larger outlier is added to a data set, the mean and median increase. When a small outlier is added to a data set, the mean and median decrease. Adding an outlier to a data set will increase the range.

This does not seem to provide an appropriate measure of center when considering this data set. What will happen if that outlier is removed? Removing the extremely large data point, 60, is going to reduce the mean to 9.3. The mean decreased dramatically because 60 was much larger than any of the other data points. What would happen with an extremely low value in a data set like this one, {12, 87, 90, 95, 98, 100}? The mean of the given set is 80. When the outlier, 12, is removed, the mean should increase and fit more closely to the other data points. Removing 12 and recalculating the mean show that this is correct. After removing the outlier, the mean is 94. So, removing a large outlier will decrease the mean while removing a small outlier will increase the mean.

Data Collection Methods

Data collection can be done through surveys, experiments, observations, and interviews. A **census** is a type of survey that is done with a whole population. Because it can be difficult to collect data for an entire population, sometimes a **sample survey** is used. In this case, one would survey only a fraction of the population and make inferences about the data. Sample surveys are not as accurate as a census, but this is an easier and less expensive method of collecting data. An **experiment** is used when a researcher wants to explain how one variable causes changes in another variable. For example, if a researcher wanted to know if a particular drug affects weight loss, he or she would choose a treatment group that would take the drug, and another group, the control group, that would not take the drug.

Special care must be taken when choosing these groups to ensure that bias is not a factor. **Bias** occurs when an outside factor influences the outcome of the research. In observational studies, the researcher does not try to influence either variable but simply observes the behavior of the subjects. Interviews are sometimes used to collect data as well. The researcher will ask questions that focus on her area of interest in order to gain insight from the participants. When gathering data through observation or interviews, it is important that the researcher is well trained so that he or she does not influence the results and the study remains reliable. A study is reliable if it can be repeated under the same conditions and the same results are obtained each time.

Making Inferences About Population Parameters Based on Sample Data

In statistics, a **population** contains all subjects being studied. For example, a population could be every student at a university or all males in the United States. A **sample** consists of a group of subjects from an entire population. A sample would be 100 students at a university or 100,000 males in the United States. **Inferential statistics** is the process of using a sample to generalize information concerning populations. **Hypothesis testing** is the actual process used when evaluating claims made about a population based on a sample.

A **statistic** is a measure obtained from a sample, and a **parameter** is a measure obtained from a population. For example, the mean SAT score of the 100 students at a university would be a statistic, and the mean SAT score of all university students would be a parameter.

The beginning stages of hypothesis testing starts with formulating a **hypothesis**, a statement made concerning a population parameter. The hypothesis may be true, or it may not be true. The test will answer that question. In each setting, there are two different types of hypotheses: the **null hypothesis**, written as H_0, and the **alternative hypothesis**, written as H_1. The null hypothesis represents verbally when there is not a difference between two parameters, and the alternative hypothesis represents verbally when there is a difference between two parameters.

Consider the following experiment: A researcher wants to see if a new brand of allergy medication has any effect on drowsiness of the patients who take the medication. He wants to know if the average hours spent sleeping per day increases. The mean for the population under study is 8 hours, so $\mu = 8$. In other words, the population parameter is μ, the mean. The null hypothesis is $\mu = 8$ and the alternative hypothesis is $\mu > 8$. When using a smaller sample of a population, the null hypothesis represents the situation when the mean remains unaffected, and the alternative hypothesis represents the situation when the mean increases. The chosen statistical test will apply the data from the sample to actually decide whether the null hypothesis should or should not be rejected.

Dependent Versus Independent Variables

Independent variables are independent, meaning they are not changed by other variables within the context of the problem. **Dependent variables** are dependent, meaning they may change depending on how other variables change in the problem. For example, in the formula for the perimeter of a fence, the length and width are the independent variables, and the perimeter is the dependent variable. The formula is shown below:

$$P = 2l + 2w$$

As the width or the length changes, the perimeter may also change. The first variables to change are the length and width, which then result in a change in perimeter. The change does not come first with the perimeter and then with length and width. When comparing these two types of variables, it is helpful to ask which variable causes the change and which variable is affected by the change.

Another formula to represent this relationship is the formula for circumference show below:

$$C = \pi \times d$$

The C represents circumference and the d represents diameter. The pi symbol is approximated by the fraction $\frac{22}{7}$, or 3.14. In this formula, the diameter of the circle is the independent variable. It is the portion of the circle that changes, which changes the circumference as a result. The circumference is the variable that is being changed by the diameter, so it is called the dependent variable. It depends on the value of the diameter.

Another place to recognize independent and dependent variables can be in experiments. A common experiment is one where the growth of a plant is tested based on the amount of sunlight it receives. Each plant in the experiment is given a different amount of sunlight, but the same amount of other nutrients like light and water. The growth of the plants is measured over a given time period and the results show how much sunlight is best for plants. In this experiment, the independent variable is the amount of sunlight that each plant receives. The dependent variable is the growth of each plant. The growth depends on the amount of sunlight, which gives reason for the distinction between independent and dependent variables.

Interpreting Probabilities Relative to Likelihood of Occurrence

Probability describes how likely it is that an event will occur. Probabilities are always a number from 0 to 1. If an event has a high likelihood of occurrence, it will have a probability close to 1. If there is only a small chance that an event will occur, the likelihood is close to 0. A fair six-sided die has one of the numbers 1, 2, 3, 4, 5, and 6 on each side. When this die is rolled there is a one in six chance that it will land on 2. This is because there are six possibilities and only one side has a 2 on it. The probability then is $\frac{1}{6}$ or 0.167.

The probability of rolling an even number from this die is three in six, or ½ or .5. This is because there are three sides on the die with even numbers (2, 4, 6), and there are six possible sides. The probability of rolling a number less than 10 is 1; since every side of the die has a number less than 6, it would be impossible to roll a number 10 or higher. On the other hand, the probability of rolling a number larger than 20 is zero. There are no numbers greater than 20 on the die, so it is certain that this will not occur, thus the probability is zero.

If a teacher says that the probability of anyone passing her final exam is 0.2, is it highly likely that anyone will pass? No, the probability of anyone passing her exam is low because 0.2 is closer to zero than to 1. If another teacher is proud that the probability of students passing his class is 0.95, how likely is it that a student will pass? It is highly likely that a student will pass because the probability, 0.95, is very close to 1.

Geometry and Measurement

Lines, Rays, Line Segments, Parallel Lines, and Perpendicular Lines

Geometric figures can be identified by matching the definition with the object. For example, a line segment is made up of two connected endpoints. A **ray** is made up of one endpoint and one extending side that goes on forever. A line has no endpoints and two sides that extend forever. These three geometric entities are shown below. What happens at A and B determines the name of each figure.

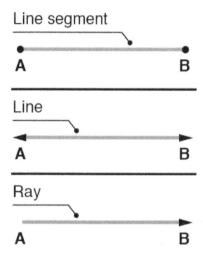

Parallel and perpendicular lines are made up of two lines, like the second figure. They are distinguished from each other by how the two lines interact. **Parallel** lines run alongside one another, but they never intersect; the distance between them always remains the same. **Perpendicular** lines intersect at a 90-degree, or a right, angle. An example of these two sets of lines is shown below. Also shown in the figure are nonexamples of these two types of lines. Because the first set of lines, in the top left corner, will eventually intersect if they continue, they are not parallel. In the second set, the lines run in the same direction and will never intersect, making them parallel. The third set, in the bottom left corner, intersect at an angle that is not right, or not 90 degrees. The fourth set is perpendicular because the lines intersect at exactly a right angle.

Lines

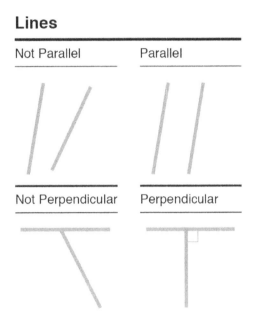

Classifying Angles Based on Their Measure

When two rays join together at their endpoints, they form an angle. Angles can be described based on their measure. An angle whose measure is ninety degrees is a right angle. Ninety degrees is a standard to which other angles are compared. If an angle is less than ninety degrees, it is an **acute** angle. If it is greater than ninety degrees, it is **obtuse**. If an angle is equal to twice a right angle, or 180 degrees, it is a **straight** angle.

Examples of these types of angles are shown below:

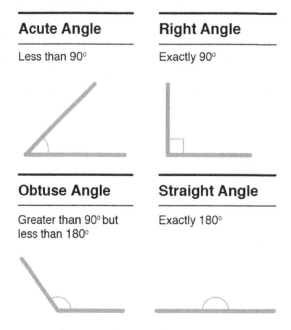

A **straight angle** is equal to 180 degrees, or a straight line. If the line continues through the **vertex**, or point where the rays meet, and does not change direction, then the angle is straight. This is shown in Figure 1 below. The second figure shows an obtuse angle. Its measure is greater than ninety degrees, but less than that of a straight angle. An estimate for its measure may be 175 degrees. Figure 3 shows acute angles. The first is just less than that of a right angle. Its measure may be estimated to be 80 degrees. The second acute angle has a measure that is much smaller, at approximately 35 degrees, but it is still classified as acute because it is between zero and 90 degrees.

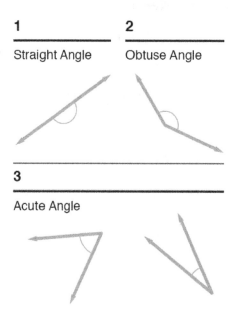

Composing and Decomposing Two- and Three-Dimensional Shapes

Basic shapes are those polygons that are made up of straight lines and angles and can be described by their number of sides and concavity. Some examples of those shapes are rectangles, triangles, hexagons, and pentagons. These shapes have identifying characteristics on their own, but they can also be decomposed into other shapes. For example, the following can be described as one hexagon, as seen in the first figure. It can also be decomposed into six equilateral triangles. The last figure shows how the hexagon can be decomposed into three rhombuses.

Decomposing a Hexagon

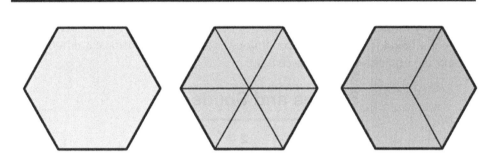

More complex shapes can be formed by combining basic shapes or lining them up side by side. Below is an example of a house. This house is one figure all together but can be decomposed into seven different shapes. The chimney is a parallelogram, and the roof is made up of two triangles. The bottom of the house is a square alongside three triangles. There are many other ways of decomposing this house. Different shapes can be used to line up together and form one larger shape. The area for the house can be calculated by finding the individual areas for each shape, then adding them all together. For this house, there would be the area of four triangles, one square, and one parallelogram. Adding these all together would result in the area of the house as a whole. Decomposing and composing shapes is commonly done with a set of tangrams. A **tangram** is a set of shapes that includes different size triangles, rectangles, and parallelograms.

A Tangram of a House

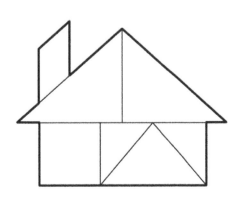

Shapes and Solids

Shapes are defined by their angles and number of sides. A shape with one continuous side, where all points on that side are equidistant from a center point is called a **circle**. A shape made with three straight line segments is a **triangle**. A shape with four sides is called a **quadrilateral**, but more specifically a square, rectangle, parallelogram, or trapezoid, depending on the interior angles. These shapes are two-dimensional and only made of straight lines and angles.

Solids can be formed by combining these shapes and forming three-dimensional figures. While two-dimensional figures have only length and height, three-dimensional figures also have depth. Examples of solids may be prisms or spheres. The four figures below have different names based on their sides and dimensions. Figure 1 is a **cone**, a three-dimensional solid formed by a circle at its base and the sides combining to one point at the top. Figure 2 is a **triangle**, a shape with two dimensions and three line segments. Figure 3 is a **cylinder** made up of two base circles and a rectangle to connect them in three dimensions. Figure 4 is an **oval** formed by one continuous line in two dimensions; it differs from a circle because not all points are equidistant from the center.

Shapes and Solids

1

2

3

4

The **cube** in Figure 5 below is a three-dimensional solid made up of squares. Figure 6 is a **rectangle** because it has four sides that intersect at right angles. More specifically, it is a square because the four sides are equal in length. Figure 7 is a **pyramid** because the bottom shape is a square and the sides are all triangles. These triangles intersect at a point above the square. Figure 8 is a **circle** because it is made up of one continuous line where the points are all equidistant from one center point.

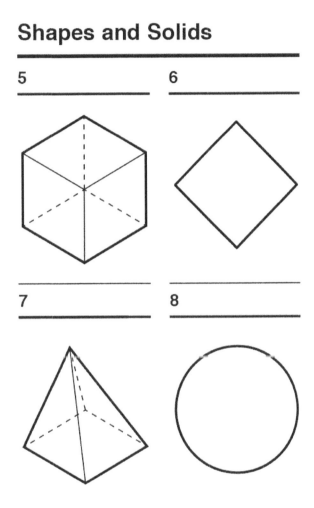

Shapes and Solids

5

6

7

8

Composition of objects is the way objects are used in conjunction with each other to form bigger, more complex shapes. For example, a rectangle and a triangle can be used together to form an arrow. Arrows can be found in many everyday scenarios but are often not seen as the composition of two different shapes. A square is a common shape, but it can also be the composition of shapes. As seen in the second figure, there are many shapes used in the making of the one square. There are five triangles that are three different sizes. There is also one square and one parallelogram used to compose this square. These shapes can be used to compose each more complex shape because they line up, side by side, to fill in the shape with no gaps. This defines composition of shapes where smaller shapes are used to make larger, more complex ones.

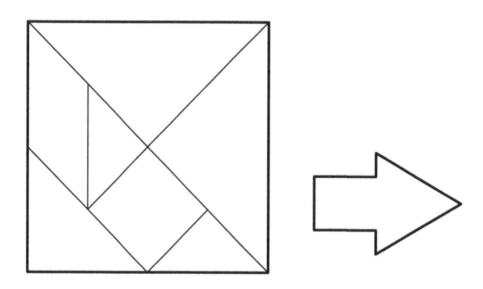

Area and Perimeter

Perimeter and area are geometric quantities that describe objects' measurements. **Perimeter** is the distance around an object. The perimeter of an object can be found by adding the lengths of all sides. Perimeter may be used in problems dealing with lengths around objects such as fences or borders. It may also be used in finding missing lengths or working backwards. If the perimeter is given, but a length is missing, subtraction can be used to find the missing length. Given a square with side length s, the formula for perimeter is $P = 4s$. Given a rectangle with length l and width w, the formula for perimeter is:

$$P = 2l + 2w$$

The perimeter of a triangle is found by adding the three side lengths, and the perimeter of a trapezoid is found by adding the four side lengths. The units for perimeter are always the original units of length, such as meters, inches, miles, etc. When discussing a circle, the distance around the object is referred to as its **circumference**, not perimeter. The formula for the circumference of a circle is $C = 2\pi r$, where r represents the radius of the circle. This formula can also be written as $C = d\pi$, where d represents the diameter of the circle.

Area is the two-dimensional space covered by an object. These problems may include the area of a rectangle, a yard, or a wall to be painted. Finding the area may require a simple formula or multiple formulas used together. The units for area are square units, such as square meters, square inches, and square miles. Given a square with side length s, the formula for its area is $A = s^2$.

Formulas for common shapes are shown below:

Shape	Formula	Graphic
Rectangle	$Area = length \times width$	
Triangle	$Area = \dfrac{1}{2} \times base \times height$	height · base
Circle	$Area = \pi \times radius^2$	radius

The following formula, not as widely used as those shown above, but very important, is the area of a trapezoid:

Area of a Trapezoid

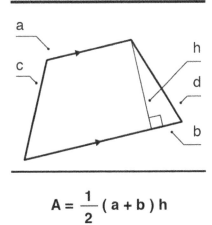

$$A = \frac{1}{2}(a+b)h$$

To find the area of the shapes above, use the given dimensions of the shape in the formula. Complex shapes might require more than one formula. To find the area of the figure below, break the figure into two shapes. The rectangle's dimensions are 6 cm by 7 cm. The triangle has a base of 4 cm and a height of 6 cm. Plug the dimensions into the rectangle formula: $A = 6 \times 7$. Multiplication yields an area of 42 cm². The triangle's area can be found using the formula:

$$A = \frac{1}{2} \times 4 \times 6$$

Multiplication yields an area of 12 cm². Add the two areas to find the total area of the figure, which is 54 cm².

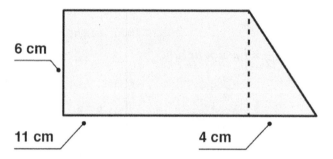

Instead of combining areas, some problems may require subtracting them, or finding the difference.

To find the area of the shaded region in the figure below, determine the area of the whole figure. Then subtract the area of the circle from the whole.

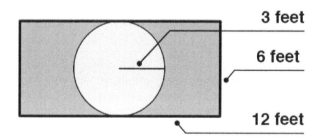

The following formula shows the area of the outside rectangle:

$$A = 12 \times 6 = 72 \text{ ft}^2$$

The area of the inside circle can be found by the following formula:

$$A = \pi(3)^2 = 9\pi = 28.3 \text{ ft}^2$$

As the shaded area is outside the circle, the area for the circle can be subtracted from the area of the rectangle to yield an area of 43.7 ft².

Geometric figures may be shown as pictures or described in words. If a rectangular playing field with dimensions 95 meters long by 50 meters wide is measured for perimeter, the distance around the field must be found. The perimeter includes two lengths and two widths to measure the entire outside of the field. This quantity can be calculated using the following equation:

$$P = 2(95) + 2(50) = 290 \text{ m}$$

150

The distance around the field is 290 meters.

Volume

Perimeter and area are two-dimensional descriptions; volume is three-dimensional. Volume describes the amount of space that an object occupies, but it differs from area because it has three dimensions instead of two. The units for volume are cubic units, such as cubic meters, cubic inches, and cubic miles. Volume can be found by using formulas for common objects such as cylinders and boxes.

The following chart shows a formula and diagram for the volume of two objects:

Shape	Formula	Diagram
Rectangular Prism (box)	$V = length \times width \times height$	length, height, width
Cylinder	$V = \pi \times radius^2 \times height$	radius, height

Volume formulas of these two objects are derived by finding the area of the bottom two-dimensional shape, such as the circle or rectangle, and then multiplying times the height of the three-dimensional shape. Other volume formulas include the volume of a cube with side length s:

$$V = s^3$$

the volume of a sphere with radius r:

$$V = \frac{4}{3}\pi r^3$$

and the volume of a cone with radius r and height h:

$$V = \frac{1}{3}\pi r^2 h$$

If a soda can has a height of 5 inches and a radius on the top of 1.5 inches, the volume can be found using one of the given formulas. A soda can is a cylinder. Knowing the given dimensions, the formula can be completed as follows:

$$V = \pi(radius)^2 \times height$$

$$\pi(1.5 \text{ in})^2 \times 5 \text{ in} = 35.325 \text{ in}^3$$

Notice that the units for volume are inches cubed because it refers to the number of cubic inches required to fill the can.

With any geometric calculations, it's important to determine what dimensions are given and what quantities the problem is asking for. If a connection can be made between them, the answer can be found.

Other geometric quantities can include angles inside a triangle. The sum of the measures of any triangle's three angles is 180 degrees. Therefore, if only two angles are known, the third can be found by subtracting the sum of the two known quantities from 180. Two angles that add up to 90 degrees are known as complementary angles. For example, angles measuring 72 and 18 degrees are complementary. Finally, two angles that add up to 180 degrees are known as supplementary angles. To find the supplement of an angle, subtract the given angle from 180 degrees. For example, the supplement of an angle that is 50 degrees is $180 - 50 = 130$ degrees.

These terms involving angles can be seen in many types of word problems. For example, consider the following problem: The measure of an angle is 60 degrees less than two times the measure of its complement. What is the angle's measure? To solve this, let x be the unknown angle. Therefore, its complement is $90 - x$. The problem gives that:

$$x = 2(90 - x) - 60$$

To solve for x, distribute the 2, and collect like terms. This process results in:

$$x = 120 - 2x$$

Then, use the addition property to add $2x$ to both sides to obtain $3x = 120$. Finally, use the multiplication properties of equality to divide both sides by 3 to get $x = 40$. Therefore, the angle measures 40 degrees. Also, its complement measures 50 degrees.

Right rectangular prisms are those prisms in which all sides are rectangles, and all angles are right, or equal, to 90 degrees. The volume for these objects can be found by multiplying the length by the width by the height. The formula is $V = lwh$.

For the following prism, the volume formula is:

$$V = 6\frac{1}{2} \times 3 \times 9$$

When dealing with fractional edge lengths, it is helpful to convert the length to an improper fraction. The length $6\frac{1}{2}$ cm becomes $\frac{13}{2}$ cm. Then the formula becomes:

$$V = \frac{13}{2} \times 3 \times 9$$

$$\frac{13}{2} \times \frac{3}{1} \times \frac{9}{1} = \frac{351}{2}$$

This value for volume is better understood when turned into a mixed number, which would be $175\frac{1}{2}$ cm^3.

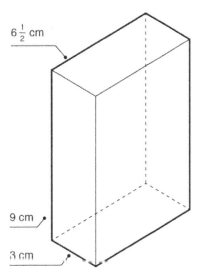

The surface area can be found for the same prism above by breaking down the figure into basic shapes. These shapes are rectangles, made up of the two bases, two sides, and the front and back. The formula for surface area adds the areas for each of these shapes in the following equation:

$$SA = 6\frac{1}{2} \times 3 + 6\frac{1}{2} \times 3 + 3 \times 9 + 3 \times 9 + 6\frac{1}{2} * 9 + 6\frac{1}{2} \times 9$$

This formula can be simplified by combining groups that are the same. Each set of numbers is used twice, to represent areas for the opposite sides of the prism. The formula can be simplified to:

$$SA = 2\left(6\frac{1}{2} \times 3\right) + 2(3 \times 9) + 2\left(6\frac{1}{2} \times 9\right)$$

$$2\left(\frac{13}{2} \times 3\right) + 2(27) + 2\left(\frac{13}{2} \times 9\right)$$

$$2\left(\frac{39}{2}\right) + 54 + 2\left(\frac{117}{2}\right)$$

$$39 + 54 + 117$$

$$210 \text{ cm}^2$$

Surface Area

Surface area is defined as the area of the surface of a figure. A **pyramid** has a surface made up of four triangles and one square. To calculate the surface area of a pyramid, the areas of each individual shape are calculated. Then the areas are added together. This method of decomposing the shape into two-dimensional figures to find area, then adding the areas, can be used to find surface area for any figure. Once these measurements are found, the area is described with square units. For example, the following figure shows a rectangular prism. The figure beside it shows the rectangular prism broken down into two-dimensional shapes, or rectangles. The area of each rectangle can be calculated by multiplying the length by the width. The area for the six rectangles can be represented by the following expression:

$$5 \times 6 + 5 \times 10 + 5 \times 6 + 6 \times 10 + 5 \times 10 + 6 \times 10$$

The total for all these areas added together is 280 cm^2, or 280 centimeters squared. This measurement represents the surface area because it is the area of all six surfaces of the rectangular prism.

The Net of a Rectangular Prism

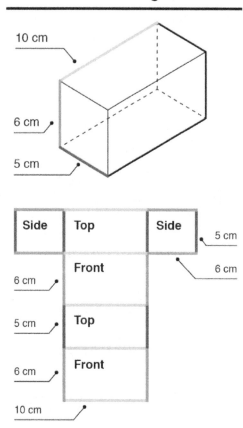

Another shape that has a surface area is a cylinder. The shapes used to make up the **cylinder** are two circles and a rectangle wrapped around between the two circles. A common example of a cylinder is a can. The two circles that make up the bases are obvious shapes. The rectangle can be more difficult to see, but the label on a can will help illustrate it. When the label is removed from a can and laid flat, the shape is a rectangle. When the areas for each shape are needed, there will be two formulas. The first is the area for the circles on the bases. This area is given by the formula $A = \pi r^2$.

There will be two of these areas. Then the area of the rectangle must be determined. The width of the rectangle is equal to the height of the can, h. The length of the rectangle is equal to the circumference of the base circle, $2\pi r$. The area for the rectangle can be found by using the formula $A = 2\pi r \times h$. By adding the two areas for the bases and the area of the rectangle, the surface area of the cylinder can be found, described in units squared.

Circles

The formula for area of a circle is $A = \pi r^2$ and therefore, formula for area of a sector is $\pi r^2 \frac{A}{360}$, a fraction of the entire area of the circle. If the radius of a circle and arc length is known, the central angle measurement in degrees can be found by using the formula $\frac{360 \times arc\ leng}{2\pi r}$. If the desired central angle measurement is in radians, the formula for the central angle measurement is much simpler as $\frac{arc\ lengt}{r}$.

A **chord** of a circle is a straight-line segment that connects any two points on a circle. The line segment does not have to travel through the center, as the diameter does. Also, note that the chord stops at the circumference of the circle. If it did not stop and extended toward infinity, it would be known as a **secant line**. The following shows a diagram of a circle with a chord shown by the dotted line. The radius is r and the central angle is A.

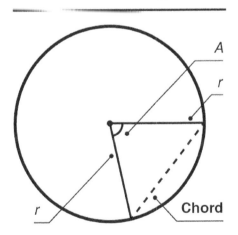

A Circle with a Chord

$$\text{Chord Length} = 2\,r\,\sin\frac{A}{2}$$

One formula for chord length can be seen in the diagram and is equal to $2r \sin\frac{A}{2}$, where A is the central angle. Another formula for chord length is: $chord\ length = 2\sqrt{r^2 - D^2}$, where D is equal to the distance from the chord to the center of the circle. This formula is basically a version of the Pythagorean Theorem.

Formulas for chord lengths vary based on what type of information is known. If the radius and central angle are known, the first formula listed above should be used by plugging the radius and angle in directly. If the radius and the distance from the center to the chord are known, the second formula listed previously should be used.

Many theorems exist between arc lengths, angle measures, chord lengths, and areas of sectors. For instance, when two chords intersect in a circle, the product of the lengths of the individual line segments are equal. For instance, in the following diagram, $A \times B = C \times D$:

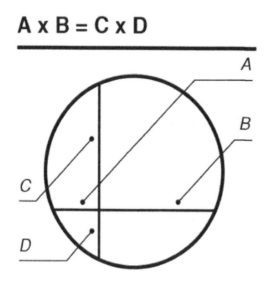

Pythagorean Theorem

Within right triangles, trigonometric ratios can be defined for the acute angle within the triangle. Consider the following right triangle. The side across from the right angle is known as the **hypotenuse**, the acute angle being discussed is labeled θ, the side across from the acute angle is known as the **opposite side**, and the other side is known as the **adjacent side**.

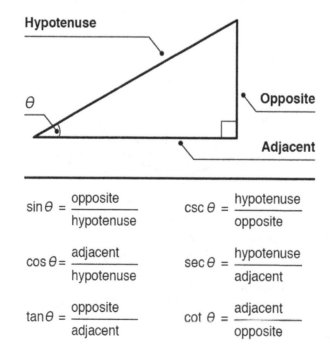

$$\sin\theta = \frac{\text{opposite}}{\text{hypotenuse}} \qquad \csc\theta = \frac{\text{hypotenuse}}{\text{opposite}}$$

$$\cos\theta = \frac{\text{adjacent}}{\text{hypotenuse}} \qquad \sec\theta = \frac{\text{hypotenuse}}{\text{adjacent}}$$

$$\tan\theta = \frac{\text{opposite}}{\text{adjacent}} \qquad \cot\theta = \frac{\text{adjacent}}{\text{opposite}}$$

The six trigonometric ratios are shown above as well. "Sin" is short for sine, "cos" is short for cosine, "tan" is short for tangent, "csc" is short for cosecant, "sec" is short for secant, and "cot" is short for cotangent. A mnemonic device exists that is helpful to remember the ratios. SOHCAHTOA stands for $Sine = \frac{Opposite}{Hypotenuse}$, $Cosine = \frac{Adjacent}{Hypotenuse}$, and $Tangent = \frac{Opposite}{Adjacent}$. The other three trigonometric ratios are reciprocals of sine, cosine, and tangent because $\csc\theta = \frac{1}{\sin\theta}$, $\sec\theta = \frac{1}{\cos\theta}$, and $\cot\theta = \frac{1}{\tan\theta}$.

The **Pythagorean Theorem** expresses an important relationship between the three sides of a right triangle. It states that the square of the hypotenuse is equal to the sum of the squares of the other two sides. When using the Pythagorean Theorem, the hypotenuse is labeled as side c, the opposite is labeled as side a, and the adjacent side is side b. The theorem can be seen in the following diagram:

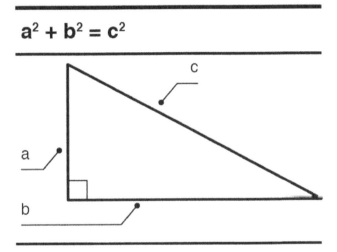

The Pythagorean Theorem

$$a^2 + b^2 = c^2$$

Both the trigonometric ratios and Pythagorean Theorem can be used in problems that involve finding either a missing side or missing angle of a right triangle. Look to see what sides and angles are given and select the correct relationship that will assist in finding the missing value. These relationships can also be used to solve application problems involving right triangles. Often, it is helpful to draw a figure to represent the problem to see what is missing.

Solving for Missing Values in Triangles, Circles, and Other Figures

Solving for missing values in shapes requires knowledge of the shape and its characteristics. For example, a triangle has three sides and three angles that add up to 180 degrees. If two angle measurements are given, the third can be calculated. For the triangle below, the one given angle has a measure of 55 degrees. The missing angle is x. The third angle is labeled with a square, which indicates a measure of 90 degrees. Because all angles must add up to 180 degrees, the following equation can be used to find the missing x-value:

$$55° + 90° + x = 180°$$

Adding the two given angles and subtracting the total from 180 gives an answer of 35 degrees.

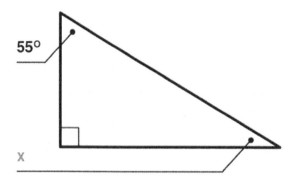

A similar problem can be solved with circles. If the radius is given but the circumference is unknown, the circumference can be calculated based on the formula $C = 2\pi r$. This example can be used in the figure below. The radius can be substituted for r in the formula. Then the circumference can be found as:

$$C = 2\pi \times 8 = 16\pi = 50.24 \text{ cm}$$

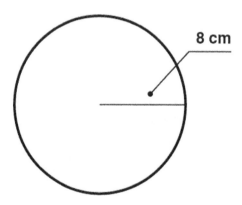

Other figures that may have missing values could be the length of a square, given the area, or the perimeter of a rectangle, given the length and width. All the missing values can be found by first identifying all the characteristics that are known about the shape, then looking for ways to connect the missing value to the given information.

Congruency and Similarity

Two figures are **congruent** if they have the same shape and same size, meaning same angle measurements and equal side lengths. Two figures are similar if they have the same angle measurements but not side lengths. Basically, angles are congruent in similar triangles and their side lengths are constant multiples of each other. Proving two shapes are similar involves showing that all angles are the same; proving two shapes are congruent involves showing that all angles are the same and that all sides are the same. If two pairs of angles are congruent in two triangles, then those triangles are similar because their third angles have to be equal due to the fact that all three angles add up to 180 degrees.

There are five main theorems that are used to prove congruence in triangles. Each theorem involves showing that different combinations of sides and angles are the same in two triangles, which proves the triangles are congruent.

The **side-side-side (SSS)** theorem states that if all sides are equal in two triangles, the triangles are congruent. The **side-angle-side (SAS)** theorem states that if two pairs of sides and the included angles are equal in two triangles then the triangles are congruent. Similarly, the **angle-side-angle (ASA)** theorem states that if two pairs of angles and the included side lengths are equal in two triangles, the triangles are similar. The **angle-angle-side (AAS)** theorem states that two triangles are congruent if they have two pairs of congruent angles and a pair of corresponding equal side lengths that are not included. Finally, the **hypotenuse-leg (HL)** theorem states that if two right triangles have equal hypotenuses and an equal pair of shorter sides, the triangles are congruent.

An important item to note is that **angle-angle-angle (AAA)** is not enough information to prove congruence because the three angles could be equal in two triangles, but their sides could be different lengths.

Similarity, Right Triangles, and Trigonometric Ratios

Within two similar triangles, corresponding side lengths are proportional, and angles are equal. In other words, regarding corresponding sides in two similar triangles, the ratio of side lengths is the same. Recall that the SAS theorem for similarity states that if an angle in one triangle is congruent to an angle in a second triangle, and the lengths of the sides in both triangles are proportional, then the triangles are similar. Also, because the ratio of two sides in two similar right triangles is the same, the trigonometric ratios in similar right triangles are always going to be equal.

If two triangles are similar, and one is a right triangle, the other is a right triangle. The definition of similarity ensures that each triangle has a 90-degree angle. In a similar sense, if two triangles are right triangles containing a pair of equal acute angles, the triangles are similar because the third pair of angles must be equal as well. However, right triangles are not necessarily always similar.

The following triangles are similar:

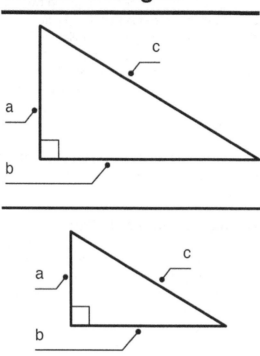

This similarity is not apparent at first glance; however, theorems can be used to show similarity. The Pythagorean Theorem can be used to find the missing side lengths in both triangles. In the larger triangle, the missing side is the hypotenuse, c. Therefore, $9^2 + 12^2 = c^2$. This equation is equivalent to $225 = c^2$, so taking the square root of both sides results in the positive root $c = 15$. In the other triangle, the Pythagorean Theorem can be used to find the missing side length b. The theorem shows that $6^2 + b^2 = 10^2$, and b is then solved for to obtain $b = 8$. The ratio of the sides in the larger triangle to the sides in the smaller triangle is the same value, 1.5. Therefore, the sides are proportional. Because they are both right triangles, they have a congruent angle. The SAS theorem for similarity can be used to show that these two triangles are similar.

Sine and Cosine of Complementary Angles

Two **complementary angles** add up to 90 degrees, or π radians. Within a right triangle, the sine of an angle is equal to the ratio of the side opposite the angle to the hypotenuse and the cosine of an angle is equal to the ratio of the side adjacent to the angle to the hypotenuse. Within a right triangle, there is a right angle and because the sum of all angles in a triangle is 180 degrees, the two other angles add up to 90 degrees, and are therefore complementary. Consider the following right triangle with angles A, B, and C, and sides a, b, and c.

Consider the following right triangle with angles A, B, and C, and sides a, b, and c.

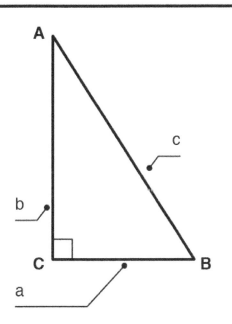

It is true by using such ratios described above that $\sin B = \frac{b}{c}$ and $\cos B = \frac{a}{c}$. Also, it is true that $\cos A = \frac{b}{c}$ and $\sin A = \frac{a}{c}$. Therefore, $\sin B = \cos A$ and $\cos B = \sin A$. A and B are complementary angles, so given two complementary angles, the sine of one equals the cosine of the other, and the cosine of one equals the sine of the other. Given the two complementary angles 30 degrees and 60 degrees, $\sin 30 = \frac{1}{2}$, $\cos 60 = \frac{1}{2}$, $\cos 30 = \frac{\sqrt{3}}{2}$, and $\sin 60 = \frac{\sqrt{3}}{2}$. Either a calculator set in degrees mode, a unit circle, or the Pythagorean Theorem could be used to find all these values.

Representing Three-Dimensional Figures with Nets

The **net of a figure** is the resulting two-dimensional shapes when a three-dimensional shape is broken down. For example, the net of a cone is shown below. The base of the cone is a circle, shown at the bottom. The rest of the cone is a quarter of a circle. The bottom is the circumference of the circle, while the top comes to a point. If the cone is cut down the side and laid out flat, these would be the resulting shapes.

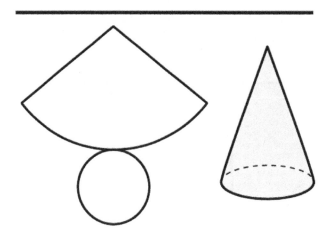

A net for a pyramid is shown in the figure below. The base of the pyramid is a square. The four shapes coming off the square are triangles. When built up together, folding the triangles to the top results in a pyramid.

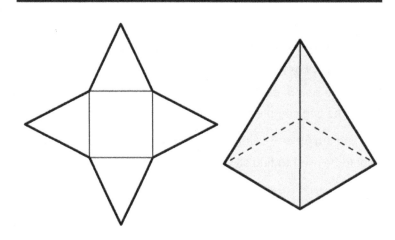

One other net for a figure is the one shown below for a cylinder. When the cylinder is broken down, the bases are circles, and the side is a rectangle wrapped around the circles. The circumference of the circle turns into the length of the rectangle.

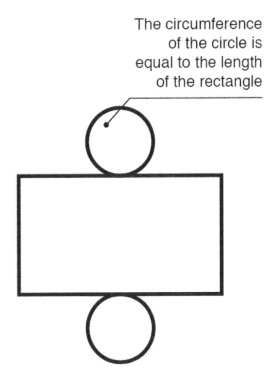

The circumference of the circle is equal to the length of the rectangle

Nets can be used in calculating different values for given shapes. One useful way to calculate surface area is to find the net of the object, then find the areas of each shape and add them together. Nets are also useful when composing or decomposing shapes and when determining connections between objects.

Using Nets to Determine the Surface Area of Three-Dimensional Figures

The surface area of a three-dimensional figure is the total area of each of the figure's faces. Because nets lay out each face of an object, they make it easier to visualize and measure surface area. The following figure shows a triangular prism. The bases are triangles and the sides are rectangles. The second figure shows the net for this triangular prism. The dimensions are labeled for each of the faces of the prism. To determine the area for the two triangles, use the following formula:

$$A = \frac{1}{2}bh = \frac{1}{2} \times 8 \times 9 = 36 \, \text{cm}^2$$

The rectangles' areas can be described by the equation:

$$A = lw = 8 \times 5 + 9 \times 5 + 10 \times 5$$

$$40 + 45 + 50 = 135 \, \text{cm}^2$$

The area for the triangles can be multiplied by two, then added to the rectangle areas to yield a total surface area of 207 cm².

A Triangular Prism and Its Net

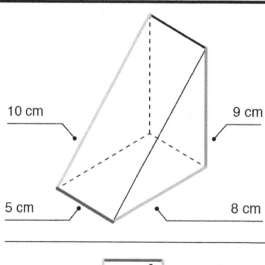

10 cm   9 cm

5 cm   8 cm

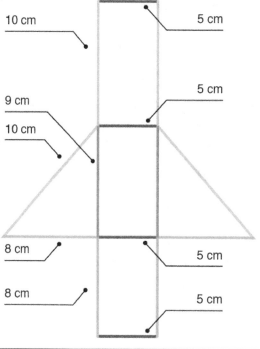

10 cm   5 cm

    5 cm

9 cm

10 cm

8 cm   5 cm

8 cm   5 cm

$$\mathbf{SA} = 2 \times (\frac{1}{2}\,bh) + lw$$
$$= 2 \times (\frac{1}{2} \times 8 \times 9) + (8 \times 5 + 9 \times 5 + 10 \times 5)$$
$$= 207\,cm^2$$

Determining How Changes to Dimensions Change Area and Volume

When the dimensions of an object change, the area and volume are subject to change also. For example, the following rectangle has an area of 98 square centimeters. If the length is increased by 2, becoming 16 cm, then the area becomes:

$$A = 7 \times 16 = 112 \text{cm}^2$$

The area increased by 14 cm, or twice the width because there were two more widths of 7 cm.

For the volume of an object, there are three dimensions. The given prism has a volume of:

$$V = 4 \times 12 \times 3 = 144 \text{ m}^3$$

If the height is increased by 3, the volume becomes:

$$V = 4 \times 12 \times 6 = 288 \text{ m}^3$$

When the height increased by 3, it doubled, which also resulted in the volume doubling. From the original, if the width was doubled, the volume would be:

$$V = 8 \times 12 \times 3 = 288 \text{ m}^3$$

The rectangle's dimensions are 6 cm by 7 cm. The same increase in volume would result if the length was doubled.

Solving Problems by Plotting Points and Drawing Polygons in the Coordinate Plane

Shapes can be plotted on the coordinate plane to identify the location of each vertex and the length of each side. The original shape is seen in the figure below in the first quadrant. The length is 6 and the width is 4. The reflection of this rectangle is in the fourth quadrant. A reflection across the y-axis can be found by determining each point's distance to the y-axis and moving it that same distance on the opposite side.

For example, the point C is located at $(2, 1)$. The reflection of this point moves to $(-2, 1)$ when reflected across the y-axis. The original point A is located at $(2, 5)$, and the reflection across the y-axis is located at $(-2, 5)$. It is evident that the reflection across the y-axis changes the sign of the x-coordinate. A reflection across the x-axis changes the sign on the y-coordinate, as seen in the reflected figure below. Other translations can be found using the coordinate plane, such as rotations around the origin and reflections across the y-axis.

Solving Problems Involving Elapsed Time, Money, Length, Volume, and Mass

To solve problems, follow these steps: Identify the variables that are known, decide which equation should be used, substitute the numbers, and solve. To solve an equation for the amount of time that has elapsed since an event, use the equation $T = L - E$ where T represents the elapsed time, L represents the later time, and E represents the earlier time. For example, the Minnesota Vikings have not appeared in the Super Bowl since 1976. If the year is now 2021, how long has it been since the Vikings were in the Super Bowl? The later time, L, is 2021, $E = 1976$ and the unknown is T. Substituting these numbers, the equation is $T = 2021 - 1976$, and so $T = 45$. It has been 45 years since the Vikings have appeared in the Super Bowl.

Questions involving total cost can be solved using the formula, $C = I + T$ where C represents the total cost, I represents the cost of the item purchased, and T represents the tax amount.

To find the length of a rectangle given the area equals 32 square inches and width equals 8 inches, the formula $A = L \times W$ can be used. Substitute 32 for A and substitute 8 for w, giving the equation $32 = L \times 8$. This equation is solved by dividing both sides by 8 to find that the length of the rectangle is 4. The formula for volume of a rectangular prism is given by the equation $V = L \times W \times H$. If the length of a rectangular juice box is 4 centimeters, the width is 2 centimeters, and the height is 8 centimeters, what is the volume of this box? Substituting in the formula we find $V = 4 \times 2 \times 8$, so the volume is 64 cubic centimeters. In a similar fashion as those previously shown, the mass of an object can be calculated given the formula, $Mass = Density \times Volume$.

Measuring and Comparing Lengths of Objects

Lengths of objects can be measured using tools such as rulers, yard sticks, meter sticks, and tape measures. Typically, a ruler measures 12 inches, or one foot. For this reason, a ruler is normally used to measure lengths smaller than or just slightly more than 12 inches. Rulers may represent centimeters instead of inches. Some rulers have inches on one side and centimeters on the other. Be sure to recognize what units you are measuring in. The standard ruler measurements are divided into units of 1 inch and normally broken down to $\frac{1}{2}, \frac{1}{4}, \frac{1}{8}$, and even $\frac{1}{16}$ of an inch for more precise measurements. If measuring in centimeters, the centimeter is likely divided into tenths.

To measure the size of a picture, for purposes of buying a frame, a ruler is helpful. If the picture is very large, a yardstick, which measures 3 feet and normally is divided into feet and inches, might be useful. Using metric units, the meter stick measures 1 meter and is divided into 100 centimeters. To measure the size of a window in a home, either a yardstick or meter stick would work. To measure the size of a room, though, a tape measure would be the easiest tool to use. Tape measures can measure up to 10 feet, 25 feet, or more.

Comparing Relative Sizes of U.S. Customary Units and Metric Units

Measuring length in United States customary units is typically done using inches, feet, yards, and miles. When converting among these units, remember that 12 inches = 1 foot, 3 feet = 1 yard, and 5,280 feet = 1 mile. Common customary units of weight are ounces and pounds. The conversion needed is 16 ounces = 1 pound. For customary units of volume ounces, cups, pints, quarts, and gallons are typically used. For conversions, use 8 ounces = 1 cup, 2 cups = 1 pint, 2 pints = 1 quart, and 4 quarts = 1 gallon.

For measuring lengths in metric units, know that 100 centimeters = 1 meter, and 1,000 meters = 1 kilometer. For metric units of measuring weights, grams and kilograms are often used. Know that 1,000 grams = 1 kilogram when making conversions. For metric measures of volume, the most common units are milliliters and liters. Remember that 1,000 milliliters = 1 liter.

Converting Within and Between Standard and Metric Systems

When working with dimensions, sometimes the given units don't match the formula, and conversions must be made. When performing operations with rational numbers, it might be helpful to round the numbers in the original problem to get a rough idea of what the answer should be. This system expands to three places above the base unit and three places below. These places correspond to prefixes that each signify a specific base of 10.

The following table shows the conversions:

kilo-	hecto-	deka-	base	deci-	centi-	milli-
1,000 times the base	100 times the base	10 times the base		1/10 times the base	1/100 times the base	1/1,000 times the base

To convert between units within the metric system, values with a base ten can be multiplied. The decimal can also be moved in the direction of the new unit by the same number of zeros on the number. For example, 3 meters is equivalent to 0.003 kilometers. The decimal moved three places (the same number of zeros for kilo-) to the left (the same direction from base to kilo-). Three meters is also equivalent to 3,000 millimeters. The decimal is moved three places to the right because the prefix milli- is three places to the right of the base unit.

The English Standard system, which is used in the United States, uses the base units of foot for length, pound for weight, and gallon for liquid volume. Conversions within the English Standard system are not as easy as those within the metric system because the former does not use a base ten model. The following table shows the conversions within this system:

Length	Weight	Capacity
1 foot (ft) = 12 inches (in) 1 yard (yd) = 3 feet 1 mile (mi) = 5,280 feet 1 mile = 1,760 yards	1 pound (lb) = 16 ounces (oz) 1 ton = 2,000 pounds	1 tablespoon (tbsp) = 3 teaspoons (tsp) 1 cup (c) = 16 tablespoons 1 cup = 8 fluid ounces (oz) 1 pint (pt) = 2 cups 1 quart (qt) = 2 pints 1 gallon (gal) = 4 quarts

When converting within the English Standard system, most calculations include a conversion to the base unit and then another to the desired unit. For example, take the following problem:

$$3 \text{ qt} = \underline{\hspace{1cm}} \text{ c}$$

There is no straight conversion from quarts to cups, so the first conversion is from quarts to pints. There are 2 pints in 1 quart, so there are 6 pints in 3 quarts. This conversion can be solved as a proportion:

$$\frac{3 \text{ qt}}{x} = \frac{1 \text{ qt}}{2 \text{ pt}}$$

It can also be observed as a ratio 2:1, expanded to 6:3. Then the 6 pints must be converted to cups. The ratio of pints to cups is 1:2, so the expanded ratio is 6:12. For 6 pints, the measurement is 12 cups.

This problem can also be set up as one set of fractions to cancel out units. It begins with the given information and cancels out matching units on top and bottom to yield the answer. Consider the following expression:

$$\frac{3 \text{ qt}}{1} \times \frac{2 \text{ pt}}{1 \text{ qt}} \times \frac{2 \text{ c}}{1 \text{ pt}}$$

It's set up so that units on the top and bottom cancel each other out:

$$\frac{3 \text{ q\cancel{t}}}{1} \times \frac{2 \text{ \cancel{pt}}}{1 \text{ \cancel{qt}}} \times \frac{2 \text{ c}}{1 \text{ \cancel{pt}}}$$

The numbers can be calculated as $3 \times 2 \times 2$ on the top and 1 on the bottom. It still yields an answer of 12 cups.

This process of setting up fractions and canceling out matching units can be used to convert between standard and metric systems. A few common equivalent conversions are:

$$2.54 \text{ cm} = 1 \text{ in}$$

$$3.28 \text{ ft} = 1 \text{ m}$$

and

$$2.205 \text{ lb} = 1 \text{ kg}$$

Writing these as fractions allows them to be used in conversions. For the problem 5 meters = ___ ft, use the feet-to-meter conversion and start with the expression $\frac{5 \text{ m}}{1} \times \frac{3.28 \text{ ft}}{1 \text{ m}}$. The "meters" will cancel each other out, leaving "feet" as the final unit. Calculating the numbers yields 16.4 feet. This problem only required two fractions. Others may require longer expressions, but the underlying rule stays the same. When a unit in the numerator of a fraction matches a unit in the denominator, then they cancel each other out. Using this logic and the conversions given above, many units can be converted between and within the different systems.

The conversion between Fahrenheit and Celsius is found in a formula:

$$°C = (°F - 32) \times \frac{5}{9}$$

For example, to convert 78 °F to Celsius, the given temperature would be entered into the formula:

$$°C = (78 - 32) \times \frac{5}{9}$$

Solving the equation, the temperature comes out to be 25.56 °C. To convert in the other direction, the formula becomes:

$$°F = °C \times \frac{9}{5} + 32$$

Remember the order of operations when calculating these conversions.

Practice Quiz

1. What is the solution to the equation $3(x + 2) = 14x - 5$?
   a. $x = 1$
   b. $x = -1$
   c. $x = 0$
   d. All real numbers

2. What is the solution to the equation $10 - 5x + 2 = 7x + 12 - 12x$?
   a. $x = 12$
   b. $x = 1$
   c. $x = 0$
   d. All real numbers

3. Which of the following is the result when solving the equation $4(x + 5) + 6 = 2(2x + 3)$?
   a. $x = 6$
   b. $x = 1$
   c. $x = 26$
   d. No solution

4. How many cases of cola can Lexi purchase if each case is $3.50 and she has $40?
   a. 10
   b. 12
   c. 11.4
   d. 11

5. Two consecutive integers exist such that the sum of three times the first and two less than the second is equal to 411. What are those integers?
   a. 103 and 104
   b. 104 and 105
   c. 102 and 103
   d. 100 and 101

See answers on next page.

Answer Explanations

1. A: First, the distributive property must be used on the left side. This results in:

$$3x + 6 = 14x - 5$$

The addition principle is then used to add 5 to both sides, and then to subtract $3x$ from both sides, resulting in $11 = 11x$. Finally, the multiplication principle is used to divide each side by 11. Therefore, $x = 1$ is the solution.

2. D: First, like terms are collected to obtain:

$$12 - 5x = -5x + 12$$

Then, the addition principle is used to move the terms with the variable, so $5x$ is added to both sides and the mathematical statement $12 = 12$ is obtained. This is always true; therefore, all real numbers satisfy the original equation.

3. D: The distributive property is used on both sides to obtain:

$$4x + 20 + 6 = 4x + 6$$

Then, like terms are collected on the left, resulting in:

$$4x + 26 = 4x + 6$$

Next, the addition principle is used to subtract $4x$ from both sides, and this results in the false statement $26 = 6$. Therefore, there is no solution.

4. D: This is a one-step, real-world application problem. The unknown quantity is the number of cases of cola to be purchased. Let x be equal to this amount. Because each case costs $3.50, the total number of cases times $3.50 must equal $40. This translates to the mathematical equation $3.5x = 40$. Divide both sides by 3.5 to obtain $x = 11.4286$, which has been rounded to four decimal places. Because cases are sold whole (the store does not sell portions of cases), and there is not enough money to purchase 12 cases, there is only enough money to purchase 11.

5. A: First, the variables have to be defined. Let x be the first integer; therefore, $x + 1$ is the second integer. This is a two-step problem. The sum of three times the first and two less than the second is translated into the following expression:

$$3x + (x + 1 - 2)$$

Set this expression equal to 411 to obtain:

$$3x + (x + 1 - 2) = 411$$

The left-hand side is simplified to obtain:

$$4x - 1 = 411$$

To solve for x, first add 1 to both sides and then divide both sides by 4 to obtain $x = 103$. The next consecutive integer is 104.

Electronics Information

Electric Charge

Electric charge is a fundamental property of matter; it is the difference between positively and negatively charged particles in a material or object. Matter is composed of atoms and molecules, which consist of three basic particles: protons, neutrons, and electrons. By convention, protons are said to possess a positive charge, while electrons have a negative charge, while neutrons possess no charge. The interactions between positively charged protons and negatively charged electrons is the basis of chemical reactions and electricity—the movement of electrons. Chemical batteries contain materials that can undergo a reaction, which results in electrons moving across a medium. Certain metals, such as copper, silver, and gold, allow electrons to flow easily due to their crystalline structure and the relatively small size of electrons. This model of electrons in metals is often referred to as a "sea" of electrons.

The main characteristic of charge is attraction and repulsion. Positive charges repel other positive charges, but positive charges attract negative charges and vice versa. In the atomic model, the electron is negative, so it is attracted to the protons in the nucleus of an atom. The repulsion of like charges and attraction of opposite charges is often simply remembered by the phrase "opposites attract."

The units for electric charge can be defined in terms of the number of electrons. For example, the charge of one electron is equivalent to -1 e, the unit for elementary charge, while $+1$ e is the charge of a proton. A more commonly used unit, the coulomb, is based on this definition. The coulomb is approximately equal to $6.242\,509\,0744 \times 10^{18}$ of these elementary charges. The variable q is commonly used to represent electric charge in equations.

Another concept important to the understanding of electricity is energy. Energy exists in multiple forms and can be described as kinetic or potential energy. **Kinetic energy** is the energy that matter possesses due to motion (a ball that has been thrown possesses an amount of kinetic energy—the energy that would be required to make it stop moving). **Potential energy** is the amount of energy stored in an object as its ability to act, whether that is a ball held in the air or a block of wood. A ball that is held in the air and let go will fall, changing its potential energy (height) to kinetic energy (motion).

A block of wood is a bunch of molecules that have formed chemical bonds. With some initial heat (a form of kinetic energy) to warm it up to a certain point, the bonds can break, and the potential energy of the bonds is then released as heat (kinetic energy). This heats up other bonds, causing them to release more potential energy as heat, and so on. The unit for energy is the **joule (J)**, which is the energy needed to move one coulomb of charge across one volt of electric potential difference.

Current

Current is the rate of flow of electrons. The basic unit of current used in physics is the **ampere (A)**, also called an **amp**, which is defined as the movement of one coulomb of charge through a point in one second. Current is typically represented by the variable I. In equation form, this is:

$$I = \frac{q}{t}$$

The variable q represents electric charge and t is time. Thus, the movement of one coulomb of charge through a given point per second is one ampere of current.

Water flow is a commonly used analogy for thinking about electrical current. Consider a tall water reservoir, held back by a dam. Opening a spillway on the dam connects one side to the other, which allows water to flow through and out of the reservoir. This is like a current in an electrical circuit—a complete electrical circuit allows electrical charges to flow from a place of high charge to one of low charge. The flow rate of water in this example is the rate at which the water flows out of the reservoir. This is similar to the rate at which charge moves through the circuit—the electrical current.

Voltage

Voltage, also known as electromotive force, has become a common term. Power sources such as batteries have a certain voltage (though in reality it varies a bit). Batteries have a certain voltage, outlets supply a certain voltage, and **Voltage** is the difference in electric potential between two points and uses the common unit of a **volt (V)** and is often represented by the variable V. The **electric potential** is defined as the amount of energy needed to move one unit of charge from one point to another. Borrowing from the previous water analogy, voltage can be thought of like water pressure: The voltage in a circuit is like the amount of pressure (voltage) that is moving water (driving charges) at a certain rate (current). In equation form, voltage is equal to the amount of energy transferred per coulomb of charge passed between two points:

$$V = \frac{W}{q}$$

Resistance

Resistance is a property of matter that resists the flow of current through it, and it is inversely proportional to conductance, the ability for current to flow through matter. Resistance is measured in units called **ohms (Ω)** and, in equations, is represented by the variable R:

$$\frac{1}{R} = G$$

Components of a circuit that use the energy in the circuit provide a certain amount of resistance. This can be considered analogous to an obstruction in the path of flowing water: if an object is placed (firmly) in a water channel, it will resist the flow of water; less water can pass through that point at one time, reducing the rate of flow. Similarly, a circuit element with resistance will resist the flow of electrons, reducing current. The wires in an electrical circuit typically have low resistance, while the insulation used around wires has high resistance, encouraging current to flow preferentially through the wire and not whatever else it might be touching. For example, air, due to its low density (as a gas), has relatively few electrons in a given space (compared to conducting materials) to allow charges to pass from one point or another, so air is characterized by very high resistance.

Resistors in circuits use a common numbering scheme to standardize the creation of resistors. The numbering system is called the **E series**, a geometric sequence designed to ensure that any given number is within a certain percent tolerance of one of its numbers. The number after E is how many numbers are in the series, between one and ten. For example, the series with the fewest items, E3, has only 3 numbers: 1, 2.2, and 4.7. Any value needed is within 40% of these three numbers. The difference between any number and the next can be calculated by using the series number as the root of ten. For the E3 series, that's $\sqrt[3]{10}$ or $10^{0.33}$, which is about 2.15. Multiplying any value in the series by the same factor gives the next value in the series. Some values used in industry deviate slightly from the calculated values, such as the numbers 4.7 and 47 (which are close to, but not quite, the exact values received from the calculation). These have become commonplace in resistors.

The resistance of a physical resistor can be found printed on it; however, many small resistors lack the space to print their values as readable numbers, so a color code system is used instead. Many resistors will have four or five bands, with the first two or three color bands representing the digits zero through nine, the next being an exponential multiplier, to allow for larger or smaller values, and the last band indicating the tolerance of the resistor (the range of variation of the resistor from its nominal value, as a percent).

Color	Number	Multiplier	Tolerance
Black	0	×1Ω	
Brown	1	×10Ω	±1%
Red	2	×100Ω	±2%
Orange	3	×1kΩ	
Yellow	4	×10kΩ	
Green	5	×100kΩ	±0.5%
Blue	6	×1MΩ	±0.25%
Violet	7	×1MΩ	±0.10%
Grey	8	×1MΩ	±0.05%
White	9	×1GΩ	
Gold		×0.1Ω	±5%
Silver		×0.01Ω	±10%

Basic Circuits

All real circuits have a voltage, current, and some amount of resistance. The simplest circuit consists of a power source (such as a battery or wall power) and a conductor (wire) to allow current to pass from one end to the other. Although this circuit wouldn't serve much purpose, it meets the requirement of a closed circuit. Despite only being a battery and wire, if this were a real circuit it would still have a (low) resistance, because all materials have a small (or even negligible) amount of resistance. Unless stated otherwise, it can be assumed that the components in diagrams (wires, power sources, and loads) are ideal and present exactly the resistance stated.

A basic functional circuit possesses parts or components (also called circuit elements) that make use of the charge being pushed through the circuit to accomplish some task, such as heating up a heating coil on an electric stovetop. The material in such a heating coil has a high melting point and high resistance, so some of the energy in the circuit is lost to the material as heat. The greater this resistance, the more energy is required to move charges across the resistance to the other end of the circuit.

One thing to understand with circuits is the convention of electrical flow. Because like charges repel each other, electrons will move toward a place with fewer electrons; thus, charges move from an area with high negative charge to an area with low negative charge. However, by convention, we say the flow of current in a circuit goes from the positive terminal of a power source toward the negative terminal. Most sources use this convention to show the flow of current, but keep in mind that the actual flow of electrons is from negative (high concentration of electrons) to positive (low concentration of electrons).

Ohm's Law

Ohm's law is the relationship between current, voltage, and resistance. It states that current is directly proportional to voltage. However, in most cases, the current is not *equal* to voltage. The factor that makes the difference is the resistance in the circuit.

To bring the water analogy together, raising the pressure (voltage) applied to the water (for example, by adding a water pump) without changing the amount of resistance in the water channel would increase the speed or flow rate (current) of the water. Increasing the amount of resistance in the water channel (wire) but keeping the flow rate the same would require a higher pressure and raising the flow rate without changing the pressure would require reducing the resistance.

The voltage at any given point in a circuit is equal to the current in that circuit multiplied by the resistance:

$$V = IR$$

Ohm's law also exists in equivalent, algebraically derived forms: $I = \frac{R}{V}$ and $R = \frac{V}{I}$. Ohm's law is particularly useful to remember when analyzing circuits. In general, if two of the quantities are known, the third can be found by Ohm's law. If the voltage and resistances in a circuit are known, the current can be calculated. Consider a simple circuit with a known voltage of 9 volts and a single 10-ohm resistor. Using Ohm's law, the current would be calculated as follows:

$$I = \frac{V}{R} = \frac{9\text{ V}}{10\text{ }\Omega} = 0.9\text{ A}$$

Thus, the current in that example would be 0.9 amps.

Series and Parallel Circuits

Series circuits connect everything along one line, from one end of the power source to the other. Parallel circuits have all their components split out onto separate paths. Most real circuits exist as series-parallel circuits, containing elements in series and parallel, but to grasp the concepts of series and parallel circuits, it's best to start simple. One method of identifying the type of circuit is that if the entire circuit can be traced with one line from start to end without overlapping, it's in series, while if it takes more than one line or overlaps, at least one part of it is parallel.

Series Circuits

Series circuits are typically the easier of the two to analyze. Most questions about series circuits ask basic questions about the quantities of the circuit, such as asking for the total resistance, or finding an unknown current or voltage. The following example shows a simple circuit:

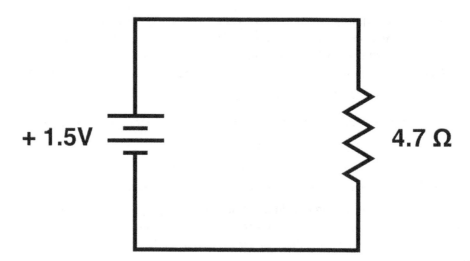

To find the unknown current, apply Ohm's law to solve for current:

$$I = \frac{V}{R}$$

Next, use the given values for voltage and resistance to arrive at the answer:

$$I = \frac{1.5 \text{ V}}{4.7 \text{ }\Omega} = 0.32 \text{ A}$$

Analyzing most circuits will involve an amount of algebra, such as rearranging Ohm's law into a more convenient form. There are a few rules to series circuits that help simplify analysis. The first rule is that the current in series is the same at all points. In the previous example, the current is 0.32 amps at the battery and on both ends of the resistor. Another rule is that the total resistance in a series circuit is equal to the sum of its resistances.

The following example shows a circuit with three resistors:

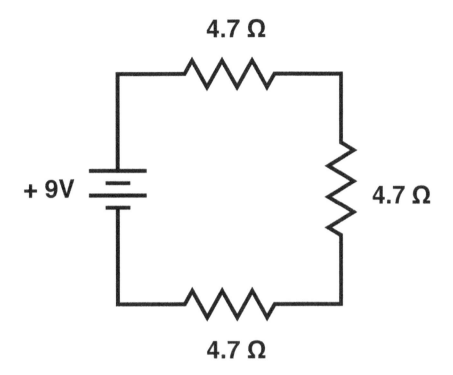

Once again, Ohm's law must be used:

$$I = \frac{V}{R}$$

To get the current for the whole circuit, however, all resistances in the circuit need to be accounted for. The battery here is 9 volts, and there are three 4.7-kiloohm resistors. Resistors in series can be added together to find the total resistance:

$$4.7 \text{ k}\Omega + 4.7 \text{ k}\Omega + 4.7 \text{ k}\Omega = 14.1 \text{ k}\Omega$$

The total resistance is 14.1 kiloohms, so the current will be the voltage, 9 volts, divided by the total resistance, 14.1 kiloohms. Remember that one ampere is one volt per ohm, so it needs to be converted by a factor of 1,000:

$$I = \frac{V}{R} = \frac{9 \text{ V}}{14.1 \text{ k}\Omega} \times \frac{1 \text{ k}\Omega}{1,000 \Omega} = 0.638 \text{ mA}$$

Thus, the current in this example is 0.638 milliamps.

If we were given the magnitude of the voltage drops across the resistors instead of the battery's voltage, it would still be possible to find the current. Consider the following circuit. The components surrounding the resistors represent voltmeters, which measure the electric potential difference (voltage) from one point to another:

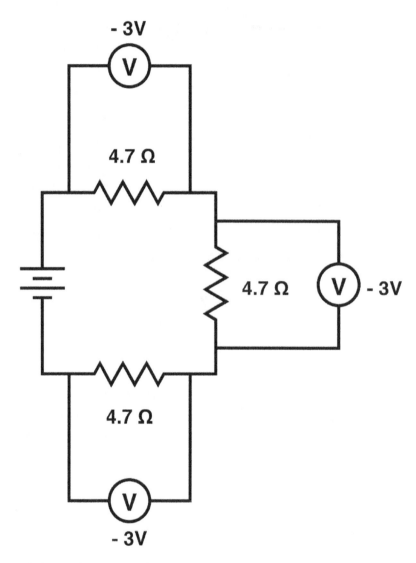

The easiest way would be to divide one of the voltage drops by the resistance it crossed, because the current will be the same at all points in the circuit. However, if only the circuit's voltage drops and total resistance were known, the total voltage applied to the circuit would be found by adding all the voltage drops. The voltage applied to the circuit is equal to the sum of the voltage drops.

Parallel Circuits

In a purely **parallel circuit**, all resistances are on separate conductors and there are effectively two points of interest in the circuit—before and after the parallel:

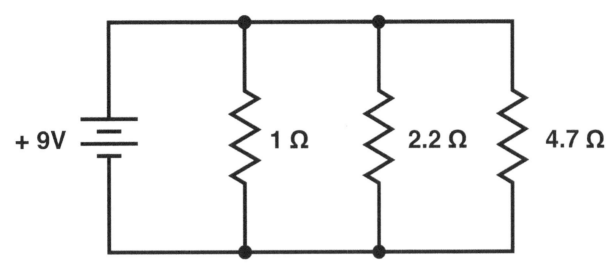

In this example, the resistances can't simply be summed together to find the total resistance—the circuit isn't in series. The following formula shows the relationship between total resistance (noted as R_{total}) and resistances in parallel:

$$\frac{1}{R_{\text{total}}} = \frac{1}{R_1} + \frac{1}{R_2} + \cdots + \frac{1}{R_n}$$

This is true for any number of resistances (n) that are in parallel. To find the total resistance then requires taking the reciprocal of both sides:

$$R_{\text{total}} = \frac{1}{\frac{1}{R_1} + \frac{1}{R_2} + \cdots + \frac{1}{R_n}}$$

The same method can be applied to the previous example circuit:

$$R_{\text{total}} = \frac{1}{\frac{1}{1\,\Omega} + \frac{1}{2.2\,\Omega} + \frac{1}{4.7\,\Omega}} = \frac{1}{1\,\Omega + 0.455\,\Omega + 0.21\,\Omega} = \frac{1}{1.67\,\Omega} \approx 0.60\,\Omega$$

Once the total resistance of resistors in parallel has been found, the diagram may be simplified to show a single resistor with the **equivalent resistance**. Note that, unlike resistances in series, the total resistance of resistances in parallel is *less* than the sum of the individual resistances:

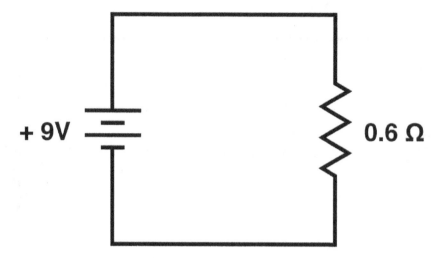

At this point, the current can be solved for the same as a simple series circuit to get 15 amps.

The rules for circuits in parallel are different from those for series circuits. To start with, the current is *not* always the same at each point in a parallel circuit. When a circuit splits into parallel branches, the current is divided, with the current through one branch inversely proportional to the resistance of that path. The following shows an example of a parallel circuit with two equal resistances:

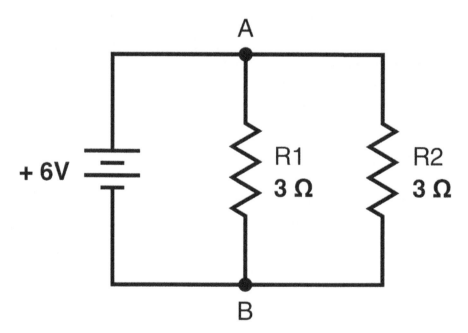

A German physicist, Gustav Kirchoff, is credited for adding to the understanding of circuits. One such realization was that the sum of currents entering a point (junction) must be equal to sum of currents exiting that same point. This is known as **Kirchhoff's current law (KCL)**. To find the current through a given resistance in a parallel circuit, you must first know the voltage at both ends of the branch. In this example, the voltage must be 6 volts from one end to the other, because that's how much the battery supplies, so current through either of the resistors is:

$$I = \frac{V}{R} = \frac{6\,V}{3\,\Omega} = 2\,A$$

According to Kirchhoff's current law, the current going into point A must be equal to the sum of currents exiting point A. Since each resistor has a current of 2 amps going through it, the total current is 4 amps going from point B to point A. This can be shown true by finding the equivalent resistance:

$$R_{total} = \frac{1}{\frac{1}{3\,\Omega} + \frac{1}{3\,\Omega}} = \frac{1}{\frac{2}{3\,\Omega}} = 1.5\,\Omega$$

Then, the current for the circuit can be calculated:

$$I = \frac{V}{R} = \frac{6\,V}{1.5\,\Omega} = 4\,A$$

Finding equivalent resistances and understanding Kirchhoff's current law can help analyzing most parallel circuits. Kirchhoff also formalized another realization about circuits—that the sum of all voltages in a loop must equal zero. This is known as **Kirchhoff's voltage law (KVL)**, which holds true for any path starting and ending at the same point, adding all voltage changes along the path. Together, the Kirchhoff's current and voltage laws are known as Kirchhoff's circuit laws.

More involved circuits contain a combination of series and parallel portions and are called series-parallel circuits or combination circuits. Consider the following series-parallel circuit:

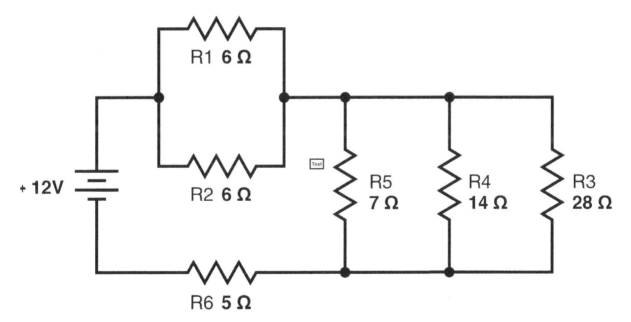

Finding the current and voltage drops in this circuit requires finding equivalent resistances for the parallel resistors. Start with the first pair of parallel resistors. We can refer to their equivalent resistance as R_A:

$$R_A = \frac{1}{\frac{1}{6\,\Omega} + \frac{1}{6\,\Omega}} = \frac{1}{\frac{2}{6\,\Omega}} = \frac{1}{\frac{1}{3\,\Omega}} = 3\,\Omega$$

Similarly, calculate the equivalent resistance (R_B) of the second set of branched resistors:

$$R_B = \frac{1}{\frac{1}{7\,\Omega} + \frac{1}{14\,\Omega} + \frac{1}{28\,\Omega}} = \frac{1}{\frac{4}{28\,\Omega} + \frac{2}{28\,\Omega} + \frac{1}{28\,\Omega}} = \frac{1}{\frac{7}{28\,\Omega}} = \frac{1}{\frac{1}{4\,\Omega}} = 4\,\Omega$$

Now that all the resistances are known, the diagram can be simplified to find the total resistance:

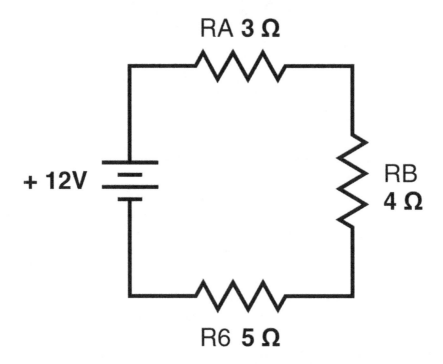

The total resistance of this circuit is simply the sum of the equivalent resistances, or 12 ohms. The total current must therefore be 1 amp. The current through any individual resistor can be found by the voltage drop across that resistor divided by its resistance. For example, consider R_4. The resistance across any resistor in parallel is the same for each other parallel resistor. So, the voltage drop across R_B must first be calculated in order to find the current at R_4:

$$V_B = I \times R_B = 1\,\text{A} \times 4\,\Omega = 4\,\text{V}$$

The current through R_4, which may be called I_4, then can be found:

$$I_4 = \frac{V_B}{R_4} = \frac{1\,\text{A} \times 4\,\Omega}{14\,\Omega} = \frac{4\,\text{V}}{14\,\Omega} = 0.286\,\text{A} = 286\,\text{mA}$$

In summary, current in parallel circuits splits across individual branches, inversely proportional to the resistance in that branch (more flow where there is less resistance), while the equivalent resistance of branches is less than their sum.

Even more complex circuits which cannot be called either series or parallel exist, but those require more involved methods to be correctly analyzed.

Electrical Power

Electrical power is the rate at which a charge moves across an electric potential difference, or the product of current and voltage. Electrical power is represented by the variable P and is relates to voltage and energy as follows:

$$P = VI = \frac{E}{q} \times \frac{q}{t} = \frac{E}{t}$$

In other words, electrical power is equal to the amount of energy used to do a certain amount of work in a given time. The unit used for power is the **watt (W)**, which is equal to one volt-ampere (one volt times one ampere):

$$1\,W = 1\,V \times 1\,A = 1\,\frac{J \times C}{C \times s} = 1\,\frac{J}{s}$$

Electrical power can also be shown in terms of resistance and current or resistance and voltage by applying Ohm's law to substitute voltage or current:

$$P = VI = I^2 R = \frac{V^2}{R}$$

Knowing the amount of power in a circuit is particularly important when the electrical energy in the circuit is being transformed into mechanical energy (e.g., by a motor).

AC Versus DC

Previous discussion has been focused on direct current (DC) circuits, where current flows steadily in one direction. However, there is another conventional type of electrical current. **Alternating current (AC)** is generated by magnetic induction. One of the properties of circuits is that electrical flow generates magnetic fields. Conversely, a magnetic field can act upon charges to induce an electrical current in conductive materials. This is known as **induction**.

The generation of alternating current is reliant upon a *moving* magnetic field, however. Simply holding a magnet near a wire will temporarily displace charges, but not keep generating a current. Alternating between positive and negative polarities however will keep pushing and pulling charges, so there *is* constantly a current, except that it periodically switches direction (therefore, *alternating* current). The rate that it alternates direction is called its frequency, which has the units of inverse seconds ("per second"), or hertz (Hz).

Capacitors

When two ends of an open circuit are brought close together (without touching), a small amount of charges builds up on the surface of the conductive material. A capacitor is a circuit component that creates a similar scenario but provides a much larger surface area than the end of a wire for a charge to be built on, effectively "charging" the capacitor. Traditionally, capacitors are constructed from parallel conductive plates separated by air or a nonconductive material. The amount of a capacitor may be charged, or its **capacitance**, is limited by the voltage used to charge the capacitor, and by the construction of the capacitor itself (the surface area of the charged plates and the properties of the matter between the plates). Capacitance is represented by the variable C and is measured using the **farad (F)** unit. When a charged capacitor is connected in a circuit with a lesser voltage, it instead generates a current. One of the properties of a capacitor is that the greater its charge, the greater its maximum current. In general, a capacitor resists changes to its voltage, which it must do by moving charges and thus creating a current.

Mathematically, the equation for capacitance is $q = C \times V$, where q is charge and V is the electric potential difference between the two charged surfaces, because it is the proportionality factor between the voltage and charge of conductors. Capacitance is thus represented by C, and in parallel plate capacitors, it may be calculated as:

$$C = \frac{q}{V} = \frac{\varepsilon_0 A}{d}$$

The surface area and distance between plates are A and d, respectively, while ε_0 (lowercase epsilon) is the constant for vacuum permittivity—the ability of an electrical field to occupy free space. The value of this constant, in farads per meter, is:

$$\varepsilon_0 = 8.854 \times 10^{-12} \frac{\mathrm{F}}{\mathrm{m}}$$

This is equivalent to 8.854 picofarads per meter, the scale many capacitors are rated in. Most things on Earth do not exist in a vacuum, however. Whether it is air or another material, the insulator between charged conductors, called a **dielectric**, needs to be accounted for:

$$C = \kappa \frac{\varepsilon_0 A}{d}$$

The constant κ (lowercase kappa) represents the dielectric constant for any material between the charged surfaces. This quantity is also known as relative permittivity, ε_r, as it is simply a factor of the permittivity of a given material compared to the permittivity of a vacuum, ε_0. The relative permittivity of a vacuum is thus, by definition, 1, because it is the same as ε_0. Air, while not exactly 1, is *relatively close* to 1 (1.00054), compared to most common dielectrics.

One aspect of capacitors is that there is a finite amount of charge they may hold. When the charge in a capacitor increases beyond its limit, charges begin to instead flow through the dielectric instead of building up against it. This phenomenon is known as **dielectric breakdown**, and the capability of a dielectric to resist this is known as **dielectric strength**, which is measured in volts per meter—the electric potential difference required before charge will flow across one meter of the dielectric.

Inductors

Inductors are components which take advantage of magnetic induction to store energy as a magnetic field. Inductors consist of wire in a coil, which results in a directional magnetic field, either around air or a magnetic core (to store more energy). Like the capacitance of capacitors, the capability of inductors to store energy is known as their **inductance**, which is represented by the variable L and the unit of henries (**H**). An inductor additionally takes some time to rise to or fall from its maximum current because it resists changes to current. In contrast to capacitors, inductors resist changes to current by creating an electric potential difference against the flow of current.

Electrical **transformers** are an important application of inductors to increase or decrease voltage by a specific ratio, determined by the physical winding of coils around a shared magnetic core. The movement of magnetic fields pushes electrons and pulls electrons, inducing an electric current.

The following equation relates the electromotive force to magnetic flux, ϕ, and time using the derivative of the electromotive force, ε (not to be confused with vacuum permittivity):

$$\varepsilon = \frac{d\phi}{dt}$$

In simple terms, the amount of EMF is proportional to the rate of change in magnetic flux over time.

Electrical Diagrams

There are some commonly used symbols used to represent different electrical components in diagrams. Wires are simply represented by a line, while most circuit elements have a specific symbol. Junctions between wires may often be represented by a dot where they intersect, otherwise, wires that overlap in the diagram but do not connect might be represented by a semi-circle to indicate their non-junction.

The following table lists some of the most common circuit elements with their symbols and equivalents.

Circuit element	Symbol	Description
Cell		A cell is contains reactants for a chemical reaction that produces electricity
Battery		A battery is multiple cells
DC power source		Current flow from a DC source is one-directional
AC power source		Direction of current from an AC source alternates
Resistor		Resistors contain materials less conducive to electric currents
Current source		A current source supplies constant current rather than constant voltage
Ground		A reference point for referring to voltage at another point (the electric potential difference between the point and ground)
Earth ground		A physical connection to the earth, allowing current to discharge into the ground
Chassis ground		A connection to a conductive or metal surface on a device
Diode		Only allows current flow in the direction of the arrow

Circuit element	Symbol	Description
Capacitor	—|(— —||—	Capacitors smooth out current and voltage by absorbing highs and releasing charge when low
Inductor	_mmm_	Inductors generate magnetic fields from direct current and resist changes to current

Real circuit diagrams will ideally show all the components in a circuit, but the symbols do not have to correspond to exact locations in a circuit if they correspond to the correct junctions and components. Most diagrams used for testing will not show the values of all components in the circuit, requiring the test-taker to remember what the symbols represent and predict or calculate unknown values from their knowledge of circuits.

Practice Quiz

1. A/an _____ is a material that contains a metallic element where charged particles called electrons can move freely.
    a. Graphite
    b. Semi-conductor
    c. Insulator
    d. Conductor

2. Which electrical component is represented by the symbol below?

    a. Battery
    b. Fuse
    c. Resistor
    d. Switch

3. Which electrical component will have the least resistance for a high frequency alternating current?
    a. Variable resistor
    b. Inductor
    c. Capacitor
    d. Fixed resistor

4. If a circuit contains three resistors of 2 Ω, 6 Ω, and 8 Ω, in parallel, what is the effective resistance?
    a. 1.5 Ω
    b. 3.2 Ω
    c. 2.4 Ω
    d. 1.3 Ω

5. Calculate the current in a circuit that contains three 5-ohm resistors wired in parallel to a 12-volt source.
    a. 7.2 A
    b. 6.2 A
    c. 5.2 A
    d. 4.2 A

See answers on next page.

Answer Explanations

1. D: Choice *D* is the correct answer. Metallic elements such as copper or aluminum are examples of conductors and contain electrons that can flow freely from one atom to another. Electricity is a natural phenomenon that results from the flow of charged particles, such as electrons. Choice *A* is incorrect since graphite is composed of carbon and acts as a semi-conductor. Choice *B* is incorrect since semi-conductors are made up of semi-metallic elements, such as silicon. Elements such as sulfur act as insulators and do not allow the flow of electricity.

2. B: Choice *B* is correct. The symbols below are labeled with the corresponding term.

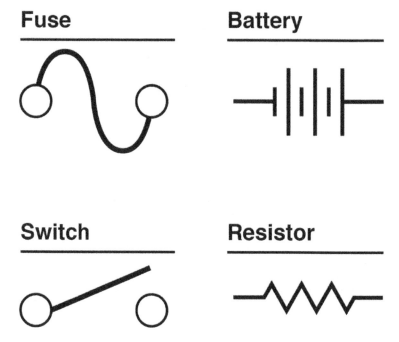

3. C: Choice *C* is correct. An alternating current (AC) can flow across a capacitor, whereas a direct current is blocked by a capacitor. Capacitance reactance (ohms) is the capacitor's opposition to the current flow. An inverse relationship exists between the capacitance reactance and the frequency of an AC signal. The lower the frequency, the greater the capacitance, or the more opposition, to the current flow through the capacitor. High frequencies mean less opposition or resistance by the capacitor. Variable or fixed resistors and inductors will resist alternating currents and are not the best answer choices.

4. D: Choice *D* is the correct answer. The effective resistance (R) for a circuit containing three resistors ($n = 3$) in parallel can be calculated from the equation below.

$$\frac{1}{R} = \frac{1}{R_1} + \frac{1}{R_2} + \frac{1}{R_3}$$

$$R_1 = 2\ \Omega, R_2 = 6\ \Omega, \text{ and } R_3 = 8\ \Omega$$

$$\frac{1}{R} = \frac{1}{2} + \frac{1}{6} + \frac{1}{8} = 0.5 + 0.167 + 0.125 = 0.792$$

$$R = 0.792^{-1} = 1.26\ \Omega \approx 1.3\ \Omega$$

5. A: Choice *A* is correct. Use Ohm's law to calculate the current:

$$V = IR$$

$$V: voltage\ in\ volts, I: current\ in\ amps, R: resistance\ in\ ohms$$

Rearrange the equation to calculate the current:

$$I = \frac{V}{R} = \frac{12\ volts}{R}$$

The resistance, or effective resistance, in parallel is:

$$\frac{1}{R} = \frac{1}{5} + \frac{1}{5} + \frac{1}{5} = 0.6$$

$$R = 0.6^{-1} \approx 1.67\ ohms$$

When the resistance is plugged into Ohm's law, the current is:

$$I = \frac{V}{R} = \frac{12\ volts}{1.67\ ohms} \approx 7.2\ amps$$

Auto Information

Types of Automotive Tools

There are a multitude of tools and measuring devices used in automotive work. In many cases, multiple tools could be used on a simple task, but there are usually tools specifically designed to perform an action more efficiently (thus, less physical effort for the user) and with less risk of damaging hardware.

Screwdrivers

The most widely used tool in automotive work is the **screwdriver**. Almost everything is assembled or fastened with some number of screws, which may be with various sizes and drive types, requiring a different screwdriver. The most encountered types of screwdrivers are Phillips (a cross or plus shape), slot ("flathead"), hex or Allen, and Torx (star) drives. Electric screwdrivers usually have swappable drive bits.

Wrenches

Almost as ubiquitous as the screwdriver is the wrench. **Wrenches** are also used for turning screws and bolts. Wrenches typically act on the exterior of a fastener head, as opposed to screws which apply torque on the interior cavity. They are especially used for bolts with no drive cavity. The variety of fastener heads also necessitates a wide range of wrench types and sizes.

Ratchet Wrenches and Sockets

Some wrenches have a ratcheting function—they only apply torque in one direction. This allows the user to move the wrench back and forth without needing to set the wrench back on the fastener head, which is particularly helpful for work in tight spaces. Most **ratcheting (socket) wrenches** have interchangeable sockets, which are usually available in sets. Breaker bars are socket wrenches with especially long handles to provide enough torque to remove stuck nuts and bolts.

Pliers

Pliers are used for gripping objects. Some of the most common are slip joint pliers, which can be opened up to a wider diameter. Another type are vise grips, or clamping pliers, which function like hand-held clamps and can be used to hold things together or out of the way. These have adjustable jaw widths to get the width just right to lock onto hardware.

Dead Blow Hammer/Mallet

A **dead blow hammer** is designed to absorb the shock of a blow. They may have internal weights that apply the force of a blow over time to help reduce bouncing after striking a surface and a plastic coating to avoid scratches. These can be used to hammer out dents or stuck bolts without causing damage.

Car Jacks and Stands

Although a full-size car lift provides more access under a vehicle, a **car jack** is a portable tool that can lift the weight of a vehicle off the ground to allow for basic maintenance such as tire replacement or changing oil. The type of jack needed depends on the weight of the vehicle, since jacks have specific weight ratings indicating the amount they can safely raise. A vehicle that's been lifted by a jack should be supported by jack stands, which have different heights and weight ratings. Aside from ratings, there are a variety of different types of car jacks used, such as scissor jacks, trolley jacks, and bottle jacks. The wider the base of the jack, the more stable it will be for lifting.

When working under a vehicle on a jack, a car creeper (or mechanic's creeper) is used to easily maneuver underneath it. Car creepers are flat panels on wheels to allow rolling, rather than trying to slide along the floor.

Air Compressor

When the tire pressure is too low, it affects the steering and tire wear. An **air compressor** is a pump that pressurizes air, which can fill up the air in a vehicle's tires with the right hose. Even without visible damage, air can leak from tires for a multitude of reasons, so they'll need to be refilled to be at optimal tire pressure.

Jumper Cables/Jump Starter

In some circumstances, a vehicle battery may have drained far enough that it's unable to start the vehicle's engine. **Insulated jumper cables** are used to recharge one vehicle's battery by using another to get it started. When using jumper cables, the order they're connected in matters, to limit the risk of igniting flammable gas from the lead-acid battery. The positive terminal of the drained battery is clamped first, then connected to the positive terminal of the working battery. Next, the negative terminal of the working battery is connected to an unpainted metal part away from the battery on the vehicle being charged, to serve as ground. The last connection made has a chance of sparking, and any flammable gas from the battery could cause an explosion. The current continues from the chassis on the vehicle back to the negative terminal via a heavy wire called the grounding strap, completing the circuit.

Starting the engine of the vehicle with the working battery will then start charging the drained battery. Once the drained battery is recharged enough to start its engine, it will be able to recharge via its alternator. The clamps are simply removed in reverse order they were put on: remove clamp from the metal chassis, then from the negative clamp on the good battery, followed by the good battery's positive terminal and finally the (formerly) dead battery's positive terminal.

A **jump starter** is a device designed to restart an engine without another vehicle available. Jump starters are effectively portable batteries with built-in cables and clips.

Cleaning Sprays and Silicone Lubricant

Dealing with most things under the hood of a vehicle involves some amount of grime or mess. **Silicone sprays are lubricants** that leave minimal residue and excess can easily be cleared away with a degreasing spray.

Safety Tools: Gloves, Goggles

Basic safety precautions when working with automobiles includes wearing gloves and safety glasses or goggles. Although not necessary for all work, gloves should be worn to prevent cuts from sharp metal edges, and when working under the engine, goggles ought to be worn since most auto fluids can irritate or damage the eyes.

Wire Strippers and Crimpers

Wire strippers are used to remove insulation from cables and expose the conductive metal underneath to be able to be able to connect and repair broken wires or add a new terminal end. **Wire crimpers** are similar to pliers. Ends of cables are added using wire crimpers, which clamp down over a wire and connector, pinching them together.

Multimeter

A **multimeter** is a must-have tool for analyzing the battery or any electrical connections. Although a modern multimeter has broader uses in electronics, the multimeter can be used to measure voltage and test continuity through a wire by attaching the positive and negative of the meter in parallel with test probes to the section being checked.

Automotive Systems

Engine System

Automotive engines are complex machines requiring many components working synchronously to optimally convert energy into power. Internal combustion engines are the primary type of automotive engine outside of electric engines, which use AC or DC motors powered by batteries. Internal combustion engines encompass both gasoline and diesel burning engines, which function on spark ignition and compression ignition systems, respectively. Spark ignition engines, however, account for the majority of internal combustion engines in the United States.

An internal combustion engine (ICE) consists of a few main components. This includes the engine block that houses the main parts of the engine; the compression chambers, which consist of cylinders around pistons, valves, and a sparkplug to ignite fuel and move the piston; the crankshaft, which is turned (or "cranked") by the action of the pistons; the camshafts, which control the timing of the intake and exhaust valves.

Cooling System

The engine's **cooling system** is responsible for keeping the engine near its optimal operating temperature (near 200 °F for most vehicles), to maximize fuel efficiency and minimize emissions. The liquid coolant used in modern vehicles is typically a mixture of water and antifreeze, which lowers the freezing point and raises the boiling point compared to pure water. This helps keep the cooling system in the liquid state.

Most modern vehicles use liquid cooling systems, which should be familiar. A liquid cooling system includes the radiator, cooling fan, coolant pump, and passageways around the engine for coolant to flow. The radiator is a metal or plastic container located in front of the engine and behind the grille. It absorbs heat from the circulating coolant and is itself cooled down by air flow. A cooling fan behind the radiator pulls air through the gaps of the radiator to cool it off.

From the radiator, coolant flows from the radiator to a coolant pump through rubber radiator hoses. This adds pressure to the system coolant in the system, both raising the boiling point of the coolant and forcing it to cycle in one direction, toward the engine. Automotive liquid cooling circulates water around the engine block via a water jacket, an enclosed space around the engine to circulate cooled fluid to absorb heat from the engine. Once the coolant has circulated through the jacket it returns to the radiator where it will be cooled down and cycle again. Some coolant systems also include a coolant reservoir. During high-temperature conditions (such as the engine running), coolant expands some and the excess flows to the coolant reservoir, where it rests until the temperature and coolant levels drop.

Fuel System

The automotive **fuel system** primarily transfers fuel from the vehicle's fuel tank (or tanks) to the cylinder chamber in the engine. In vehicles with internal combustion engines, gasoline begins at the fuel tank and is moved toward the engine by means of a fuel pump. The fuel pump in modern internal combustion engines is electric, rather than mechanical. The advent of electronic fuel injection in the '70s and '80s

192

eliminated the need for carburetors, which were previously used in conjunction with mechanical fuel pumps to mix gasoline with oxygen to achieve better fuel-to-air ratios, improving combustion efficiency.

From the fuel pump, gasoline is transferred into the fuel lines. Automotive fuel lines are made of either metal or special synthetic materials which are insoluble to gasoline. The fuel lines then move gasoline to the vehicle's fuel filter. The fuel filter on vehicles prevents dirt particles and debris from moving forward by means of a fine porous material called the filtration medium, which varies depending on the required size of the holes in the filtration medium. Smaller holes catch more debris, but also create more restriction on fuel flow. The average pore size of filtration media is measured in microns (micrometers, μm), a thousandth of a millimeter, and thus they catch most unwanted particles in the fuel line before they can clog the small interiors of the electronic fuel injection.

From the fuel filter, it moves to the fuel injector. Depending on the engine, pressurized fuel is sprayed as a fine mist into an inlet manifold where it mixes with air, while other modern vehicles use direct fuel injection, which sprays straight into the combustion chamber.

Lubrication System

The engine's **lubrication system** is what supplies oil to all the moving parts, to reduce friction and avoid overheating. When not in use, oil rests at the bottom of the engine in the oil pan. The Society of Automotive Engineers (SAE) grades motor oil numerically by its viscosity with a two-part system. For example, consider SAE 5W–30 oil, a commonly used motor oil. The first number, 5, (followed by W) refers to its viscosity in cold weather, while the second number, 30, references its viscosity while hot at 100 °C. If motor oil is too thick, it is harder to pump to the engine, while oil that is too thin can degrade (burn) and fail to properly lubricate the engine.

The oil pump is what moves oil from the oil pan up to the moving parts of the engine. Most vehicles feature mechanical oil pumps which are connected to the engine's crankshaft by a gear train to reduce the speed of the oil pump. When the crankshaft begins to rotate, the oil pump actuates and creates suction, pulling oil from the pan through an oil strainer that blocks larger particles in the bottom of the oil pan into a pickup tube through the oil filter and to the rest of the engine via a feed line, where it primarily lubricates the crankshaft, but also sends oil to lubricate the camshaft and, in engines with them, the rocker arms and push rods.

Ignition System

The **ignition system** is what enables the vehicle to get the engine started. In a modern ignition system, there are two circuits that ultimately take energy from the battery to create a high voltage spark in the combustion chambers of the engine. These are referred to as the primary and secondary ignition systems.

The primary ignition system includes the battery, the ignition switch, the ignition control module, and the ignition coil. Turning the vehicle key in the ignition completes the circuit through the ignition switch, allowing current to flow from the battery, through the switch, to the control module, and ultimately into the primary windings of the ignition coil. The return path from the ignition coil is back to the control module. The control module is responsible for timing the ignition circuit to be in sync with the engine's pistons.

The secondary ignition system includes the secondary windings of the ignition coil, a high-tension cable called the coil wire (designed to endure high voltages), the distributor, which includes its housing, cap, and rotor, and the spark plug wires and spark plugs. The secondary windings in the ignition coil are tighter, with many turns, so they greatly increase voltage, usually up to 100 times the battery's original voltage. The high voltage is passed on to the distributor through the coil wire, a resilient conductor with thick insulation to prevent unwanted sparks. The distributor is responsible for sending the high voltage current to the spark plugs so that they fire in the correct sequence at the right time. A timing circuit with a reluctor (a rotating, toothed ring triggering a magnetic switch) ensures the timing for the distributor. Finally, the actual ignition occurs when the high voltage current reaches the spark plug. The large voltage is able to overcome the dielectric strength of air inside the combustion cylinder to create a brief electric arc (spark) at the tip of the plug, which is what ignites the compressed fuel-air mixture injected by the fuel system.

Electrical System

The **electrical system** of a vehicle consists of the battery and all wires and electrical components, which includes the ignition, fluid pumps, lights, and connects to some part of most of the vehicle's systems, since it provides electricity.

Several systems' electrical components require the engine to be running to operate, except for those that might be used when the engine isn't running (such as the lights or radio). This is typically accomplished by these systems being connected past the vehicle's ignition switch, so they receive power when the switch is activated. The ignition switch may have multiple positions allowing different groups of electronics to be active with or without the ignition.

The **car battery** is the initial component of the electrical system and the source of its power. The type found in most combustion engine vehicles is a lead-acid battery. The standard twelve-volt car battery typically contains six two-volt wet cells, which use lead plates immersed in sulfuric acid as their diodes and electrolyte. As the battery discharges, the chemical reaction is driven forward until there is little difference between the diodes and the battery's acid is mostly dilute.

The typical car battery does not have a large enough capacity to run a vehicle for long periods of time, because it is limited by the amount of reactants in its cells, so it needs a way to receive energy to run the reaction in reverse, or be recharged. The vehicle's alternator is the part responsible for recharging the battery. An alternator converts part of the energy created by the engine into an electrical current to provide energy to overcome the reverse reaction's activation energy.

The alternator is kept in check by its voltage regulator. The regulator acts as a switch that turns the alternator on and off depending on output voltage of the battery. This prevents overcharging the battery, which can create excessive heat, evaporating off the water in the battery, both limiting the battery's effectiveness (due to a smaller area immersed in the electrolyte solution) and greatly increasing risk of damage to the battery (by melting or even explosion). Consequently, the voltage regulator is critical to maintaining the battery's lifespan.

Transmission System

The **transmission system** converts the power from the engine into rotation of the driveshaft while ensuring that the engine maintains a near-optimal speed. The transmission system as a whole encompasses all components from the engine's input shaft to the drive shaft and axles. There are a couple factors that affect the basic design of the car's transmission system—whether the vehicle is manual or automatic drive, and whether it has front-, rear-, or four-wheel drive.

The transmission includes the clutch and gearbox which converts the high-speed rotations of the engine's crankshaft into slower, powerful rotations in the driveshaft.

The moving parts in an automatic transmission require lubrication much like the engine, but the transmission also uses it as a hydraulic fluid, so the transmission fluid in a car differs from the oil used for lubricating engines.

Depending on whether the vehicle uses a front-, rear-, or four-wheel drive, it will have a different layout in its transmission, but several transmission components are essentially the same. The drive shaft connects the transmission to the wheels' axles via an assembly of gears called the differential, which allows the left and right wheels to rotate at different rates, because during turns the outer wheels need to cover more distance than the inner wheels.

Brake System

An automotive braking system is responsible for slowing a vehicle or bringing it to a full stop. There are several types of **braking systems** currently found in modern vehicles.

The most common braking systems use hydraulic brakes. **Hydraulic brakes** use brake fluids to transmit force from the brake pedal to the brake calipers. The basic components of a hydraulic brake system include the brake pedal, master cylinder, brake lines and fluid, and, depending on the type of hydraulic brakes, brake calipers and rotors, or brake shoes and drums.

The **master cylinder** in the brake system is designed to amplify the force from the brake pedal with a piston and cylinder, which compresses the brake fluid in the brake lines, thus applying force on the brake calipers or shoes. The master cylinder contains a reservoir of additional brake fluid that makes up for the change in volume as the brake pads and rotor are worn down, requiring the calipers to push further for the pads and rotors to contact. A spring in the master cylinder acts against the brake pedal to return it to its original position when not in use.

Disc brakes are a type of friction brake which uses a (mostly) incompressible fluid to transfer force from the brake pedal to the brakes as pressure to create friction between the stationary and rotating surfaces. In a disc brake, the part moving with the wheel is called the rotor, also known as the brake disc. The kinetic energy of the vehicle is thus lost as heat due to friction between the rotor and brake pads. Drum brakes use a brake drum and shoes rather than disc and brake pads. The force is applied outward toward the rotating drum, instead of perpendicularly inward against a rotating disc. These are less frequently used than disc brakes and, when used, are more common on rear wheels.

Brake fluid used in vehicles is usually a liquid made from glycol ethers. Brake fluid is largely non-compressible, so any force applied against it will be transmitted against its container—or in the case of the brake system, the brakes. Although most liquids are (mostly) non-compressible, brake fluids specifically need to be resistant to both high and low temperatures to remain in a liquid phase at a fixed viscosity. If viscosity increased (the fluid thickened), the brakes would become less responsive. Moisture in the air can gradually absorb into glycol-ether based brake fluids, which reduces the boiling point of the brake fluid and the efficacy of the brakes, eventually requiring fluid replacement. Other brakes use silicone or mineral oil-based fluids, which have the advantage of not absorbing atmospheric moisture and have a higher boiling point as well.

Although mainly found in vehicles with electric motors, another type of brake is the electromagnetic brake. These function by running a current through a coil to generate a magnetic field, which moves an armature against the brake. The armature is analogous to the brake pads in a hydraulic brake, and the friction material serves the same role as the brake rotor. This friction material is located between the armature and the coil, which is also called the brake field, and encases the brake field.

195

Exhaust System

The **exhaust system** in a vehicle provides an exit path for exhaust fumes from the engine. The parts that make up this system include the cylinder head (leading exhaust away from the ignition cylinders), exhaust manifold, oxygen sensor, catalytic converter, exhaust pipe, muffler, and tailpipe.

After the power stroke in the engine cycle, most fuel in the cylinder should be combusted. At this point, the gas and air mixture will have burned into a combination of steam, nitrogen gas, carbon dioxide, and other byproducts, collectively referred to as the vehicle's emissions. This gas leaves the cylinder during the exhaust stroke and enters the vehicle's exhaust manifold, which combines the exhaust from all the engine's cylinders to a single exhaust pipe. The exact shape and size of the exhaust manifold and exhaust pipe affect the pressure of the exhaust, which matters because increased pressure in the exhaust system can reduce efficiency of the engine.

The oxygen sensor in an exhaust system is used to optimize the fuel to air ratios in the engine. Depending on the amount of air coming from the cylinders, data from the oxygen sensor is used by the engine control unit to adjust the fuel injected into the cylinders; more air in the exhaust indicates there's room for more fuel injection.

Although the main products of combustion are relatively harmless, incomplete combustion yields gaseous byproducts that may be harmful to the environment. Since no engine is completely efficient, some amount of fuel may exit the cylinder without being burned completely. Because of this, vehicles produced in the last few decades feature catalytic converters which use chemically inert precious metals such as platinum to reduce the amount of nitrous oxides and carbon monoxide emitted.

Suspension System

Automotive **suspension systems** are designed to lessen the impact from surface imperfections in the road, such as bumps, short dips, and holes. The suspension allows the wheels to move up and down in relation to most of the vehicle, absorbing energy and limiting the perceived impact. This system includes the tires, shocks, and struts which support the vehicle's weight.

As part of the suspension, the tires are the interface between the road and the vehicle and the first part that absorbs bumps. The composition and wear of the tire as well as the amount of air affects how well it grips the road and how well it resists sudden vertical acceleration, affecting the vehicle's performance.

There are a few types of suspension systems that may be encountered in vehicles today, which are categorized as either dependent or independent suspension systems. The difference between dependent and independent suspensions is that **independent suspension systems** enable wheels to raise and lower separately (independently) from the others, while **dependent suspension systems** have the vertical movement of the left and right wheels linked together along the axle, so they alternately rise and fall in unison.

Shock absorbers connect from the wheels of a vehicle to its frame, and as their name suggests, their main role is to absorb vibrations. Shock absorbers may be integrated with a spring or be separate from the spring.

In front-engine, front-wheel drive vehicles it's common to find shock absorbers integrated with springs, called MacPherson struts, which is a type of independent suspension. These connect to the wheel by a wishbone-shaped arm, giving them the name wishbone suspension. The relative simplicity of this device means minimal production cost and fewer moving parts to wear down, also making it a type that can be simply replaced. A variation on this is the double-wishbone suspension, which connects by an upper and

196

lower arm and provides more control but at an increased cost and more difficult maintenance. Even more complex is what's known as a multi-link suspension, with several arms that attach laterally and longitudinally to the wheel. These provide the most control but are expensive systems with significantly more joints that become difficult to repair.

One type of spring used is the coil spring, among the most common in automotive suspensions. This is the same kind of spring also commonly used in other applications, except for vehicles that are significantly thicker and able to support the weight of a vehicle. The coil spring attaches at one end to the wheel and the other end to the frame.

Another type of shock absorber is called the leaf spring. The **leaf spring** uses multiple strips of curved metal made to different lengths and aligned so that the middle is thicker than the ends. The center of the "leaf" is connected to the axle, while the ends of the leaf support the frame, arcing upward. When the wheel is forced toward the body, the leaf spring is resistant to straighten, becoming progressively more resistant the further it is deformed. Variations of this spring in suspensions may even use two leaves facing each other in an elliptical arrangement, or one leaf spring and one half of a leaf spring connected to each other on one end and the frame on the other.

Automotive Components

Although most structural components of a vehicle are designed to last, there are several parts that need regular maintenance or replacement to keep the vehicle running. These can be found across most of a vehicle's systems.

Battery
The vehicle's **battery** has a nominal (listed) voltage, but the voltage depends on its charge. For the lead acid batteries used in gasoline vehicles, they are not designed to be fully discharged and are instead kept topped up by the alternator. As they wear, the electrolyte levels in the battery can decrease by water evaporation, terminals can become corroded by hydrogen gas, and the plates in the wet cells themselves can be degraded, reducing the effective surface area for electric transfer. All of these can reduce the battery's performance.

Electrical Cables and Connectors
The **electrical cables** in the vehicle are susceptible to some wear over time. The wires connecting the battery to the ignition and ancillary systems are critical but can become corroded starting from the terminal and creeping past the insulation. Corrosion increases electrical resistance and thus creates additional heat, which can damage the insulation and contribute to a short, which can quickly drain the battery.

Lights
Automotive lights are simple components to replace. For most vehicles, the headlight bulbs are nearly as simple to replace as home lightbulbs—many can simply be pulled out of the housing, unplugged from the electrical connector, and swapped for a new one. One must take care to keep the bulb and electrical connections clean of dirt and oil from one's hands, which will reduce the effective lifespan of a new light.

Tires
Tires are one of the most regularly worn parts of the vehicle. The effectiveness of the tire is affected by its tread depth, and the thinner the tread gets, the more likely the tires will slip or skid and the longer it will take to brake to a stop. A tread depth tool can tell more precisely the depth of the tire's tread (in

millimeters or fractions of an inch) but in a pinch, a U.S. quarter can be inserted head down into the tread to approximate the tread depth. If the head is partly covered, the tread is above 4/32", while if the whole head is visible, it is less than or equal to that thickness.

Suspension Components

Shocks and Struts

The **suspension components** need replacement every 50,000 miles. The suspension system takes the brunt of the impact from bumps in the road, gradually degrading the springs and increasing the wear in the shocks' hydraulic pistons, causing leaks in hydraulic fluid. Coil springs are simple components that can only be replaced once they're worn. Sealed shock absorbers must also be replaced once at the end of their lifespan, while others (e.g., MacPherson struts) can be repaired by swapping out just the shock absorber.

Filters

Several **filters** in the vehicle also require regular checks. As a filter gradually gets dirtier, the flow rate or quality of the filtered fluid (the filtrate) may be reduced.

The **fuel filter** is located along the fuel line from the tank to the engine. One can tell if it needs replacement by checking the color of the filter paper—if it is dark and brown, rather than yellow or tan, it has built up particulate matter on the outside of the filter and needs replacement.

The **oil filter** is located near the bottom of the engine and filters out the dirt and grime built up in the engine oil from the engine's body, combustion products on the cylinder walls, and even degradation of the oil itself. The color change on oil filters also indicates their quality. Though not a guaranteed measure of oil effectiveness, the darker the oil gets, the more contaminated it is. When the vehicle's oil is changed, replacing the filter at the same time prevents contaminating the new oil.

The engine's air intake is filtered and requires occasional replacement, around every 15,000 to 30,000 miles, or more depending on the filter manufacturer. Once the filter has gone from yellow or tan to black, that's an indication that the filter no longer can allow enough air in for optimal fuel-to-air ratios.

Fluids

As the oil cycles through the engine, it picks up most loose particles on surfaces and eventually needs replacement to remain effective. Heated oil can even be burned away, causing the oil level to decrease. Because the engine oil keeps parts lubricated to prevent friction and heat rapidly degrading parts, it's important that it is replaced when needed.

The general procedure for changing out engine oil is as follows: After safely raising the vehicle, any cover over the oil pan needs to be removed and the pan checked for oil leaks. Next, the cap may be removed to let the old oil flow out by gravity into a funnel and drain pan. Old oil may have worn the oil filter, so taking it out and replacing with a fresh filter keeps new oil clean longer—otherwise, residues captured by an old filter can potentially circulate back through the engine. At this point, the engine can be refilled with oil.

Brake Components

For disc brakes, the pads that grip the brake rotor are gradually worn thin by applying the brakes. The thinner the pads get, the less well they dissipate heat. When the pads are completely worn out, the calipers holding the brake pads can damage the brake rotor. Brakes can sometimes wear out evenly, and one side's brake pads may require replacement before the other. However, replacing brake pads should often be done in pairs, to keep equal brake effectiveness on each side of the vehicle.

Procedures for Automotive Troubleshooting and Repair

Short of dismantling a vehicle, there are telltale sights, smells, and sounds that can be used to diagnose automotive problems. The first step, however, is to check the vehicle's computer for any recorded issues.

The **engine control unit (ECU)** can provide more information in the form of error or fault codes. Some codes can be specific to the make and model of the vehicle by using internal sensors. The ECU error codes (also called diagnostic trouble codes, DTCs) were standardized by the Society of Automotive Engineers (SAE) and published as SAE J2012. These give more detail than the vehicle's dashboard lights tell the user, indicating the specific problem and its location in the vehicle. Scanners can be used to read the ECU's memory for recent errors.

The **error codes** are 5 characters long, starting with a letter followed by 4 digits. An error code starting with "P" indicates the sensed issue is part of the powertrain, which includes the engine and transmission systems. Error codes starting with "C" indicate the issue is part of the chassis, such as the brake or suspension systems. "B" errors are part of the body, which includes many ancillary systems and components in the passenger compartment. "U" error codes are related directly to the electrical system and the vehicle's computer. The first digit is either a "0" which indicates that the error is an SAE code, or a "1" for an extended code designated by the manufacturer. The second digit indicates which part of the vehicle's system has an error, and the remaining two digits indicate the specific problem. Occasionally, the final digit may be a capital letter ("P250C", for example).

Once the fault has been identified in the computer, it's time to look at the component in question. The error codes are useful diagnostics but will not always give away what caused the problem in the first place, which could result in repeat occurrences. The most significant problems tend to involve the powertrain or chassis.

The following table shows some examples of common diagnostic codes and what they indicate. Most of the possible errors for a vehicle are related to the powertrain.

SAE Codes	Definitions
P0171-P0175	Oxygen sensor issues.
P0420, P0430	Poor performance in catalytic converter.
P2300-P2335	Ignition coil issues.
P0457	Loose or missing fuel cap.
P040-P040F	Exhaust Gas Recirculation (EGR) valve issues.
P0300-P0314	Engine misfires.
P0440-P045F	Evaporation system (EVAP) issues.

The most common repairs will require having at least some tools handy: a ratchet wrench and various sockets, one or multiple screwdrivers for different fastener heads (or one with interchangeable driver bits), pliers or clamps of various sizes, and a portable light to see.

Common Automotive Issues

Squealing or Grating Sound While Braking

A squeaking or grinding sound while braking means the brake pad is worn and is rubbing against the brake rotor. To replace the brakes, the vehicle must be jacked up and the wheel must be removed. Afterwards, the brake assembly will be accessible. After removing the slider bolts and pivoting the caliper, the old brake pads and retaining clips can be replaced. Once that is finished, all parts can be readjusted and reassembled. The process can be repeated for the other side.

Clunking Noise While Turning

A knocking sound while executing a turn means the ball joints need to be replaced. To replace a ball joint, the vehicle must be jacked up and the wheel must be removed. The brake assembly may have to be adjusted or removed to access the ball joint. Next, remove the cotter pin and unfasten the castellated nut. After loosening the ball joint, the control arm can be adjusted to remove it. Carefully install the new ball joint, apply the appropriate amount of torque, and grease the assembly. Once that is finished, all parts can be readjusted or reassembled. The process can be repeated for the other side.

"Hot" Reading on the Temperature Gauge or Greenish Colored Puddle Under Vehicle

If the temperature gauge has a reading near or in the red, it may be a sign of a coolant leak. **Coolant**, or antifreeze, typically has a greenish yellow color and a sweet odor. If antifreeze is leaking from a pipe or radiator, smoke will start to form once it hits the hot engine. One symptom of this type of leak is little to no antifreeze in the coolant vat. The steps involved in fixing a coolant leak depend on where the leak is happening. The source can usually be found directly above the location of the center of the puddle.

Pink or Red Colored Puddle Under Vehicle

Transmission fluid is thick and sticky and can be pink, red, brown, or black in color depending on its usage. Negligence in fixing a transmission fluid leak in a timely manner can lead to costly repairs or replacements. Typically, a transmission leak is caused by a damaged pan gasket. While replacing a pan gasket, be sure to appropriately tighten the bolts in the pan. Failure to do so may result in another leak or further damage to the vehicle. If the leak is not caused by the pan gasket, fixing the leak will usually involve expensive replacements to the transmission system.

Dark Brown Colored Puddle Under Vehicle

A dark brown or colorless puddle may be a symptom of a brake fluid leak. A **brake fluid leak** can be a serious and immediate danger to the vehicle operator and those nearby. To check for a brake fluid leak, apply pressure to the brake pedal while the vehicle is off. If fluid is found leaking near one of the wheels or under the master cylinder, it is most likely a brake fluid leak. The steps involved in fixing a brake fluid leak depend on where the leak is happening. Leaks near the wheel may require a removal and reinstallation of the brake assembly.

Black Colored Puddle Under Vehicle

A dark, thick puddle that is located near the front of the vehicle is a sign of a leak in the oil pan. It may also have a rainbow-like appearance in the sun. An oil leak is a very serious issue that requires immediate attention to mitigate expensive damage to the engine and keep the operator safe. Any leaks in the oil pan require the oil pan to be replaced. If there is a hole in the oil pan, then it must be replaced. First, remove the mounting bolts that are located on the oil pan. Once that is finished, the oil pan can be removed from the engine block, and the mounting surface on the engine can be cleaned. Afterwards, the new oil pan and gasket can be installed. Be sure to use the appropriate sequence and amount of torque when reapplying the mounting bolts.

Water Dripping from Tailpipe

Water dripping from a vehicle's tailpipe is a natural occurrence due to weather, humidity, and other factors. Normal operation of the vehicle removes water from the exhaust system and prevents rust and damage, as long as the vehicle is running for more than a few minutes and a few miles at a time.

Vehicle Lift

A **vehicle lift** is a necessity when doing extensive work on the underside of vehicles. The lift is stationary equipment used to raise the entire vehicle and hold it in place while workers stand underneath. There are a few types of automotive lifts commonly used, though the principal is the same for all of them: getting the vehicle off the ground and keeping it stable. The main difference between most lifts is the raising mechanism. A lift may have two posts or four, use a scissoring mechanism, be built in-ground, or instead have a ramp to be driven up on. The ideal lift to use is based on space available and what it's needed for.

Because incorrectly using an automotive lift presents a risk to the worker, vehicle, and surrounding area, learning how to use the lift is crucial. With a **two-post lift**, there's an additional step to lifting a vehicle—finding its center of gravity. The engine's weight often leaves this in the front of a car, under the steering wheel or driver's seat, depending on if it has front- or rear-wheel drive. It's important to note that the points used to lift the vehicle must be on the frame of the vehicle and not the body or some other part not intended to support the weight of the vehicle.

Some vehicles may have jack pads built into the bottom of the frame, which are flat points specifically intended to use for lifting the vehicle. Wide parts of the frame rail also make for sturdy lift points, as long as balance has been considered. Other good lift points include differentials, which can support the weight of the vehicle or contact points on an axle. If in doubt, always refer to the vehicle manufacturer's instructions for its lift points, because making a mistake with a lift can be both expensive and dangerous. The adapter connected to the lift arm needs to be selected correctly for the lift point being used. Some vehicle manufacturers may also recommend that specific adapters be used.

After a vehicle's lift points have been settled, the lift controls themselves are simple. The lift arms can be raised or lowered together using the lift's control panel, usually located on the passenger side for a two-post lift. Many are simply up and down buttons to adjust height. For safety, the operator should never try to shift or push a vehicle on a lift, nor allow people inside the vehicle.

Practice Quiz

1. Vehicles equipped with a four-wheel-drive use a _____ to transfer power to the axles located in the rear and front end.
    a. Torque converter
    b. Transmission
    c. Transfer case
    d. Power steering case

2. When a battery discharges electrical energy, what happens to the lead plates in the battery?
    a. They are converted to a metal composite.
    b. They break down and release oxygen gas.
    c. They are converted to basic salt.
    d. They release electrons and are converted to lead (II) sulfate.

3. Which stroke cycle of the piston is depicted below?

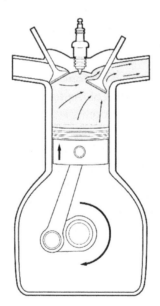

    a. Intake
    b. Compression
    c. Combustion/power
    d. Exhaust

4. Which of the following engine component facilitates compression of air and fuel in the engine combustion chamber?

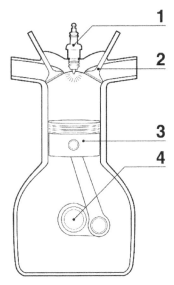

- a.   1
- b.   2
- c.   3
- d.   4

5. Which air-fuel mixture would be considered lean?
- a.   17:1
- b.   14.7:1
- c.   13:1
- d.   10:1

See answers on next page.

Answer Explanations

1. C: Choice *C* is correct. The transfer case is a vital metal component of the vehicle's drivetrain and is responsible for transferring power. That power moves from the transmission to the rear or front axles through the drive shafts. Choice *A*, the torque converter, is incorrect and is a dynamic component that transfers rotating power from the engine to a load or transmission. Choice *B*, the transmission, is a type of gearbox that uses gears/trains to convert torque/speed from the engine (rotating power source) to the drivetrain. Choice *D*, the power steering case, is incorrect and is a gearbox that uses high-pressure fluid to help turn or convert the rotary motion of the steering wheel.

2. D: Choice *D*, formation of lead (II) sulfate, is correct. A car battery contains a negative and positive electrode. The general reaction consists of the reaction of lead, lead (II) oxide, and sulfuric acid. The products are lead (II) sulfate and water. During the discharge process at the negative electrode, solid lead will lose two electrons (oxidation). Lead (II) ion is formed, which then reacts with a sulfate anion to produce lead (II) sulfate. A similar reaction occurs at the positive electrode. Lead (II) oxide reacts with sulfuric acid and gains two electrons (reduction) to produce lead (II) ion and water. Lead (II) ion reacts with a sulfate anion to produce lead (II) sulfate.

3. D: Choice *D*, the exhaust cycle, is the correct choice. The individual strokes for a four-stroke cycle engine are depicted below.

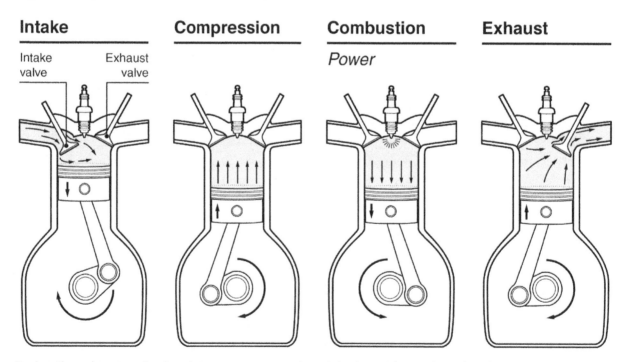

During the exhaust cycle, the piston moves upward, and the byproducts of combustion move outward through the open exhaust valve into the exhaust manifolds.

4. C: The correct answer choice is the piston head, Choice *C*. The piston head is subjected to large forces as the air/fuel mixture is compressed and heated following combustion of the mixture. Choice *A* is incorrect and refers to the spark plug that is responsible for igniting the air/fuel mixture. Choice *B* is incorrect and is an exhaust valve that opens to release the byproducts of the combustion process. Choice *D* points to the crankshaft and is responsible for converting a linear motion into a rotational motion.

5. A: Choice *A*, 17:1, is the best choice. An air-to-fuel ratio compares the air mass to the fuel that it's mixed with. Choice *B*, 14.7:1, is the ideal stoichiometric air-to-fuel ratio for a gasoline-fed engine. There are 14.7 pounds of air combined with 1 pound of fuel. Choice *A* is an example of a lean mixture where there is too much air and little fuel. These mixtures burn slowly since the distance between fuel molecules is greater and results in longer times for combustion; lean mixtures burn hotter. Choice *C* is incorrect and closer to the ideal fuel mixture, and Choice *D*, 10:1, is incorrect since it is an example of a rich air-to-fuel mixture. When a fuel is burning richly, there is not enough air and too much fuel; these rich mixtures will burn more quickly due to the shorter distance between the fuel molecules. Rich mixtures burn cooler and may result in the fouling of spark plugs followed by the production of black smoke from the exhaust.

Shop Information

Several jobs in the military, such as a Carpentry and Masonry Specialist, Construction Equipment Repairer, and Allied Trade Specialist, require a good basic understanding of shop information. A person entering the military heading into one of these types of jobs needs to have an understanding of the tools, their use them, and how to use them safely in a safe working environment. The Armed Services Vocational Aptitude Battery (ASVAB) will assess your entry-level knowledge in these areas.

Types of common shop tools and their uses

Striking Tools

- Hammers: Hammers are used for a variety of purposes such as driving nails, shaping and molding metal, or breaking materials apart. A hammer with a sharp edge, such as a claw hammer, can be used to pry or chip away at materials.

- Die and punch: A die and punch set is used to punch holes, typically into thin metal such as sheet metal. The metal is placed on the die, which has a hole in it, and the punch is driven into the hole on the die, making a hole in the metal.

- Chisels: Chisels are used (often in combination with a hammer or other striking tool) to cut into wood, stone, or metal. They have a sharp edge that cuts on impact.

- Wedges: A wedge is a triangular tool, such as the head of an axe, that is used to separate or split material. Or a wedge can lift objects by separating them from a surface.

Cutting Tools

- Power saw: A power saw is cutting tool that is mechanically powered by electricity or some other energy source. It can cut a variety of materials.

- Manual saw: A manual saw is a cutting tool that is powered by human strength. It has sharp teeth along the blade and can be used to cut a variety of materials.

- Lathe: A lathe is a mechanized tool that rapidly rotates material, such as a piece of wood, so that it can be cut symmetrically around its axis.

- Planer: A planer uses a blade to scrape away thin layers from a piece of wood to make it thinner or flatter.

- Snips: Snips are hand tools that resemble scissors and are used for essentially the same purpose. However, snips can cut through much tougher materials than scissors, such as sheet metal.

- Drill press: A drill press is a machine that drills holes in an object as the user lowers the spinning drill bit downward onto the object.

Holding Tools

- Clamp: A clamp is a tool that uses inward pressure to hold objects together. Their use can be temporary, as is often the case in carpentry, or more permanent, as is sometimes the case in plumbing.

- Pliers: Like a clamp, a pair of pliers is a hand tool designed to hold objects together with inward force. Pliers can be used to bend materials or to turn small objects like nuts and bolts with more torque than hands alone are capable of.

- Vice: A vice is a tool, usually attached to a table, that is used to securely hold an object while work is performed on it. It consists of two panels, or "jaws," one of which is movable, that are compressed together using a screw mechanism.

- Wrench: A wrench is normally used to either turn an object, such as a nut or a bolt, or to hold such an object in place. There are many styles of wrenches.

Measuring Tools

- Tape measure: A tape measure is a long strip of flexible material that is used to measure distance. It can be held in a spring-loaded mechanism, which is typical for carpentry or construction, or it can be loose, which is typical for sewing and tailoring.

- Plumb line: The plumb line, or plumb bob, consists of a string, rope, or chain with a weight fastened to the end of it. The purpose of the plumb line is to measure the vertical uprightness or horizontal incline of a structure.

- Squares and levels: A square is a tool used as a reference for a 90-degree angle. Squares come in a wide variety of designs for different uses. A level is used to determine the horizontal incline or tilt of a structure. It is read by looking into a glass tube containing liquid and a bubble; the bubble indicates the angel of the object being measured.

- Gauge: A gauge is any device used to measure a dimension. Gauges can be used to measure length, width, or depth, as well as pressure, sound levels, brightness, and many other physical aspects.

Fasteners

- Screws: Screws are threaded along a shaft so that, when they are rotated, they fasten materials together. Wood screws are used to attach wood pieces together. Masonry screws are used with concrete. Sheet metal screws connect pieces of metal. Lag screws (or lag bolts) are used with a variety of materials.

- Nails: A nail is a basic fastener with a sharp point used to drive materials together. Nails come in a large number of varieties for a number of different purposes.

- Bolts, nuts, and washers: Bolts and nuts are often used together to fasten materials between them; the nut spirals along a thread of the bolt to press the materials together. In some cases, a washer is used to distribute the pressure of the bolt.

- Rivets: Rivets are permanent fasteners that are used almost exclusively with metal. When installed, the tip expands to lock the rivet into place.

- Anchors: Anchors, or screw anchors, are small plastic tubes that are inserted into holes of materials that are too loose or brittle to hold a screw on their own. The anchor is inserted first, then the screw is screwed into the anchor. This anchor expands and creates a firm hold.

Air vs. Electric vs. Manual Tools

- Air tools: Air tools, or pneumatic tools, use compressed air to do work. There are many types, such as a nail gun, jackhammer, or rivet gun, and they can be used for many purposes. The advantage of using air tools is that they are powerful and easy to operate. Unfortunately, they require compressed air to work, so they can be less portable than other tools.

- Electric tools: Electric tools are any tools that require electricity to operate. They may be rechargeable and portable, or they may need to be attached directly to a power source at all times. They can be used many tasks such as cutting, drilling, and driving screws. The biggest benefit of electric tools is that they are capable of doing work much faster than manual tools while usually still being highly portable. The disadvantage is that they tend to be expensive and they require an electrical power source.

- Manual tools: Manual tools are tools that are powered by the human body. They can be used for cutting, striking, fastening, and many other purposes. The advantage of manual tools is that they are less expensive than electric or air powered tools and can be more easily controlled. The disadvantage of manual tools is that they take more effort to use for long periods of time than other tools.

Shop PPE and Safety

Accidents in the workplace can happen at any time. In a workshop, the presence of numerous tools makes the risk for bodily injury even greater. Having personal protective equipment (PPE) and being aware of your actions and surroundings can help lower those risks.

Goggles, Masks, Face Shields

Facial protection in a shop is vital. Safety goggles, used to protect the eyes, should be worn at all times because the potential for flying debris in the shop is high. Face masks should be worn over the mouth and nose to prevent inhalation of harmful materials, such as dust and gases. Face shields, which cover the entire facial area, should be used whenever there is risk of dangerous debris flying and hitting your face, such as sparks, metal, or even wood.

Gloves

A sturdy pair of work gloves is a good idea for hand protection, particularly while handling rough or heavy materials that can hurt, cut, or injure, such as wood or metal. However, it is important to note that the practice of wearing gloves while operating machinery can be very dangerous, as gloves can easily get caught in a machine's moving parts.

Aprons

Shop aprons can help to keep your body clean and, more importantly, protected from chemicals and abrasions. Different apron materials, such as PVC plastic, Kevlar, and leather can protect the wearer from different things. Aprons also come in different designs to better protect different parts of the body. For example, while most aprons cover the torso, some aprons include sleeves to protect the arms, while others extend down to and around each leg for lower body protection.

Proper Footwear

While footwear needs vary based on the working environment, it is always important to have comfortable footwear that prevents slipping. Many work shoes are specially made to be anti-slip, with soft, tacky rubber soles. Many styles of work boots have a steel toe for added protection. Additionally, these boots can have different heights, providing more protection to the lower leg as needed.

Ventilation System

A proper ventilation system is a must in most workshops. Any shop dealing with toxic chemicals needs to have a good ventilation system to draw away the inhalable toxins. Even something as simple as a woodworking shop needs proper ventilation for the elimination of dust and other particles in the air. Exhaust hoods and fans, as well as ductwork and filters, are necessary for a quality ventilation system.

Fire Safety

Every workplace should have preventative measures for fires and ways to handle fires in case one develops. Workshops have the potential for multiple causes of fires, including friction or sparking from tools, electrical issues, and/or flammable chemicals. As such, a variety of fire safety equipment is recommended. This can include smoke detectors, fire extinguishers, sprinkler systems, and fire blankets. Additionally, workers need to be trained in their use, as well as in operating procedures regarding what to do in case of a fire.

Practice Quiz

1. Which tool would be used to make a 1/4-inch-wide hole in a piece of wood?
   a. Die and punch
   b. Drill press
   c. Band saw
   d. Ball-peen hammer

2. Which fasteners are washers used with to distribute pressure?
   a. Nails
   b. Rivets
   c. Bolts
   d. Anchors

3. What is the name of the tool that is measuring vertical uprightness in this image?

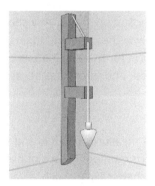

   a. Line and sinker
   b. Level
   c. Wedge
   d. Plumb line

4. Why is it unsafe to wear gloves when using a belt sander?
   a. Gloves can get caught in the moving parts.
   b. Gloves can make the user's hands slip.
   c. Gloves make it harder to operate the tools.
   d. Gloves increase the risk of carpal tunnel syndrome.

5. What are wall anchors used for?
   a. To make the heads of screws flush with the wall
   b. To ensure a better grip for nails in walls
   c. To hold a drill steady while making holes in a wall
   d. To ensure a better grip for screws in walls

See answers on next page.

Answer Explanations

1. B: A drill press is the tool that would be used to make a 1/4-inch-wide hole in a piece of wood. Choice *A* is incorrect because a die and punch system is used to put holes in thin metal. Choice *C* is incorrect because a band saw is made to cut wood, not making small holes in it. Choice *D* is incorrect because a ball-peen hammer is a striking tool, which is not used to make small holes in wood.

2. C: Washers are frequently used with bolts to distribute the pressure that a bolt and nut can produce. Choices *A*, *B*, and *D* are incorrect because they don't use washers or require pressure distribution.

3. D: The tool indicated is called a plumb line. It uses gravity to create a straight vertical line that a structure is compared to. Choice *A* is incorrect because "line and sinker" is not the name of a tool. Choice *B* is incorrect because a level uses a bubble in a glass tube to measure whether something is level. Choice *C* is incorrect because a wedge is used to separate or lift materials.

4. A: Gloves should not be worn while operating machinery because they can get caught in the machine's moving parts. Choice *B* is incorrect because gloves generally increase a user's grip. Choice *C* is incorrect because while gloves could make use of the tools more difficult, the question is about safety. Choice *D* is incorrect because work gloves have no impact on developing carpal tunnel syndrome.

5. D: A wall anchor is used to ensure a better grip for screws in a wall by expanding within the hole as the screw enters it. Choice *A* is incorrect because wall anchors don't make the heads of screws flush with a surface; the style of head on the screw does. Choice *B* is incorrect because wall anchors are not used with nails. Choice *C* is incorrect because it is the responsibility of the user to hold a drill steady during use.

Mechanical Comprehension

Physics Review and Mechanical Principles

Physics is the study of matter, energy, and the forces that interact between masses.

Scalar and Vector Quantities

A property of an object that may be described numerically is called a **quantity**. All quantities have an "amount," or their **magnitude**. In physics, many numerical quantities are used to represent objects in the real world or realistic scenarios. When an object moves, it doesn't only move at a certain speed, it moves in a specific direction as well. A numerical quantity that possesses direction is called a **vector quantity**, while a quantity without direction is called a **scalar quantity**. An example comparing a scalar and vector quantity is the difference between speed and velocity. Speed is the rate of change in distance covered over time—how quickly something moves, regardless of direction. However, this is not the same as the change in position over time, which is velocity. Because an object's position exists in more than one dimension (i.e. not on a single line), velocity is therefore also a vector. A variable representing a vector quantity may sometimes have an arrow over it to indicate it's a vector.

Vector Operations

Most of the same operations can be performed with vectors as with scalar quantities. Addition and subtraction can be solved both graphically and numerically. The graphical method is referred to as the **heads-to-tails method**, which plots vectors on a coordinate plane from end to end. Where the chain of vectors ends marks endpoint of the vector sum.

Because vectors take place in more than one dimension, they cannot be added together as simply as scalar values. If an object moves five meters west over three seconds, and then five meters east over another 3 seconds, it arrives back where it started. Its net displacement over those six seconds would be zero meters, with a distance traveled of ten meters.

The numeric method of performing vector operations separates vectors into dimensional components. To begin working with vectors, it's easiest to start with two dimensions.

To add vectors in two dimensions, the x and y components must be separated out. Assigning directions on the coordinate plane is arbitrary, but to be correct it must be consistent. By convention, an increase in y is typically treated as either the upwards or northwards direction. Consider a ball thrown straight upwards: the ball would only have a y component to its velocity because there is no horizontal movement—it goes straight up and falls back down. For two-dimensional vectors in a coordinate plane, the x and y components of each vector are added separately. A line segment can represent the magnitude and direction of a force acting upon an object.

For example, consider a person driving somewhere in a town. For every leg of their trip, they travel a certain distance and direction. If northwards is defined as positive y, then every part of their trip can be separated into x and y components. The vehicle travels 3 miles north, then takes an on-ramp and travels 10 miles southeast, and then takes an exit and travels 2 miles southwest to their destination. The first leg of the trip was 3 miles north, or $y = 3$. The second leg of the trip was 10 miles southeast. Handling diagonal movement with vectors requires the Pythagorean theorem or trigonometry to separate out the x and y components. Ten miles southeast isn't simply 10 miles south and 10 miles east, it's the hypotenuse of a right triangle, where the angles between the hypotenuse and legs are both 45°. To find the components of this vector, rearrange the Pythagorean theorem to solve for the length of the unknown legs of the triangle:

$$x^2 + y^2 = 10^2$$

Since the direction is directly southeast, both the x and y components are the same length:

$$x^2 + x^2 = 10^2$$

$$2x^2 = 10^2$$

$$x^2 = \frac{10^2}{2}$$

$$x = \sqrt{50} = 10 \times \sqrt{2} \approx 7.07$$

Thus, the south and east components are each about 7.07 miles. Using trigonometric functions and a calculator greatly simplifies the calculation. Given the angle of one leg, the length of the opposite or adjacent leg can be found by multiplying the hypotenuse (10 miles) by the sine or cosine of the angle (45 degrees), respectively:

$$\sin 45 \times 10 \text{ mi} \approx 7.07 \text{ mi}$$

Displacement, Velocity, and Acceleration

To be able to describe something's position, it must be in reference to something else. When saying something moved north, this uses the Earth and its geography as a reference point. However, there are cases where it becomes difficult to describe the position or other quantities of an object without it. If the statement is made that "the box was moved two meters to the right," there is ambiguity because it's uncertain where exactly "right" is, because it depends on the observer. This is where we make use of the **Cartesian coordinate system**. A coordinate system has three **axes** (labelled x, y, and z by convention) and their intersection is called the **origin**, which is given the value of 0 on all axes. This allows problems and systems to be described in three dimensions. When modeling a system using coordinates, assigning the origin to a point that remains in place relative to most other things is usually ideal.

Displacement is a vector quantity describing the change in position from a point of reference. This quantity is similar to distance and is measured in the same units of length (such as feet or meters), but it includes a direction, making it a vector quantity. For example, if an object is lifted six inches up, moved horizontally by two feet, then put back down, it will have moved a distance of three feet, but its displacement will only have been two feet. In other words, distance is the total length an object travels within a frame of reference, while displacement is the change in position from the start point to end point in the same reference frame.

Velocity is the change in displacement over time. This is analogous to **speed** (distance over time), but the difference is it has a direction because it is the change in *displacement* over time, making it a vector. This can be written mathematically as follows:

$$\vec{v} = \frac{\Delta x}{\Delta t}$$

In this case, the displacement is in one dimension, x. Consider a vehicle driving on the highway along a long curve at 70 miles per hour. It may have a constant speed, but its velocity is continuously changing because it's turning left or right, so the direction it's moving keeps changing. In other words, the turning vehicle is accelerating.

Acceleration is a change in velocity over time. This could be an increase or decrease in velocity of an object moving in a straight line, a change in the direction of an object's velocity, or both. Acceleration is measured in units of length over time squared, such as meters per second squared ($\frac{m}{s^2}$). For example, a ball held in place on a slope starts out moving at zero meters per second and when it's let go, it takes two seconds to increase its velocity to ten meters per second as it rolls down the slope. The ball would be said to have an average acceleration of 5 meters per second squared. In equation form, this would be:

$$a = \frac{\Delta v}{\Delta t} = \frac{\Delta x}{\Delta t \times \Delta t} = \frac{x_f - x_i}{(t_f - t_i)t} = \frac{10 \text{ m}}{(2 \text{ s})^2} =$$

One important point is that an object that is accelerating does not necessarily have to be moving. Consider again a ball thrown straight up into the air. At its highest point, its velocity is zero—for an instant when the ball is at the very top of its path, it's neither moving up nor down. However, it is still accelerating downwards due to gravity.

Newtonian Physics & Newton's Laws

Newtonian physics is the subset of physics that deals the movement of macroscopic objects. **Macroscopic objects** are those large enough to be seen by the human eye, as opposed to atoms and molecules.

Gravity is one example of a Newtonian force. Objects in space (not outer space, just *any* space) exert a force on other objects. The amount of force is proportional to the product of their masses and inversely proportional to the distance between them. This is known as **Newton's law of universal gravitation**, which closely approximates the force of gravity upon one object due to the mass of another. On the surface of Earth, the change in acceleration due to gravity is negligibly affected by altitude. Thus, gravity (on Earth) is generally considered the same everywhere, 9.81 m/s² downward.

Newton's first law of motion states that an object will keep a constant velocity unless acted upon by a force. In other words, if the net force (the vector sum of all forces) on an object is zero, an object will keep moving or remain at rest. **Inertia** represents the tendency of an object to resist changes in its velocity, and for this reason Newton's first law is also known as the law of inertia. Inertia is quantified by the concept known as **momentum**, which is the product of an object's velocity and mass. As an example, consider a car and a rubber ball rolling down a hill at the same speed. Stopping the rubber ball by hand is easy, but stopping the car requires significantly more effort. Even though they are moving at the same speed, the car has significantly more mass, and thus greater momentum.

Newton's second law of motion describes the acceleration of an object in relation to its mass and the net force acting upon it. This law is inversely proportional to the mass off the object. For example, an object at rest that is pushed with a certain force will accelerate in the same direction as the push, and the heavier that object is, the slower its velocity will change.

Newton's third law of motion is possibly the most well-known of Newton's laws. It states that for any force exerted upon an object, the object exerts an equal and opposite force. If you try to push or pull a sandbag along a floor, you are exerting a force on the bag, but at the same time, the bag exerts the exact *opposite* force against you. This can be illustrated with an example of a person jumping. When a crouched person extends their legs quickly to jump, they are pushing against the ground. However, this works because the ground simultaneously exerts a force against the jumper's feet, pushing them upwards. The ground (the whole earth, for that matter) is considerably more massive, so the effect of a person jumping is generally imperceptible. However, if a person jumps off a diving board, their downwards push is partially resisted by the diving board, and the board bends downwards before springing back and returning to its original shape.

Since all objects possess mass, an object resting on a table is pulled downward by gravity and exerts a downward force against the table. The force exerted in return by the table upon the object is called the **normal force**. In this case "normal" does not mean "regular." It refers to the direction of the force being perpendicular to the surface. If an object rests on a slight incline, the normal force is not straight upward, but rather perpendicular to the incline (friction provides the force parallel to the incline preventing its acceleration). A normal force is exerted in proportion to the mass of a given object; a heavier object has a greater mass, and since all objects are accelerated (nearly) equally by gravity, its force against the surface it is contacting is greater.

Force

In the most general sense, a force pushes or pulls an object. A **force** can be described as a change in the velocity of a given mass, or the acceleration of a mass. This is what we think of as **weight**, the acceleration due to gravity of an object of a certain mass.

Free body diagrams are models used to analyze the forces acting upon an object. To create a free body diagram, a system must be drawn out and an arrow should be drawn for every force acting upon each object. For most hypothetical scenarios, this will include the normal force of an object against a surface or another object it is in contact with.

According to Newton's third law, for every force exerted upon an object, there is an equal and opposite force.

Friction

Despite industrial attempts to minimize it, everything on Earth is subject to *some* amount of friction. **Friction** is the result of molecular scale interactions between objects in contact with each other. What may look like a perfectly flat surface is a mesh or lattice of molecules, which may be weakly attracted to molecules of another object, and the combination of *many* weak attractions creates a strong enough attractive force to resist motion. The sum of these microscopic interactions can be generalized to create a good approximation of the friction between macroscopic objects. Energy lost to friction is usually in the form of heat, like rubbing hands together in the cold—the friction generated from hands moving back and forth against each other creates heat, warming them up.

Because friction is based on the amount of surface area between objects, the amount of friction also depends on what the objects are made of. A rough surface will have less contact against a smooth one than two smooth surfaces or two rough surfaces would.

The coefficients of friction for various materials depends on the rigidity and surface smoothness, but it may also be measured experimentally. A basic method involves putting an object (of a known mass) on one end of a flat surface or ramp and raising the end with the object, then measuring the lowest angle at which the object begins to slide. The free body diagram of the object would include the downwards force due to gravity, the normal force on the object perpendicular to the ramp, and the force of friction. The force of friction will be parallel to the ramp, opposite of the direction the object slides.

The friction between two objects not moving in relation to each other is called **static friction**. The maximum force of static friction is directly proportional to the normal force upon an object. The more an object pushes against another (such as due to the weight of the object), the closer the objects are and thus the greater the friction. This is one reason it's harder to push something heavy.

Objects in motion are also under the effect of friction, but for objects in motion, it is referred to as kinetic friction. **Kinetic friction** represents the amount of force that needs to be overcome for an object to stay in motion while sliding along a surface. Once the force of kinetic friction exceeds the forces opposing it, the object becomes stationary in relation to the surface. Consider a cardboard box sliding down a ramp. The normal force in this system is perpendicular to the contact between the box and the ramp and is equal to the force of gravity on the object times the cosine of the ramp's angle. The friction between the box and the ramp is parallel to the ramp, and in the opposite direction of its motion.

The coefficient of friction is represented by the symbol mu, μ, and the coefficients of static and kinetic friction are distinguished as μ_s and μ_k.

The following table lists some common materials and the coefficients of static and kinetic friction between them. By comparison, human joints have very *low* friction. The exact values of a specific material will vary, but many such tables provide values for making approximations. Type of surface, adhesive properties, and temperature all affect friction, and thus the experimentally obtained coefficients of friction.

Interaction	Coefficient of Static Friction	Coefficient of Kinetic Friction
Wood on wood	0.25–0.5	0.2
Metal on wood	0.2–0.6	0.2
Steel on dry steel	0.5–0.8	0.42
Steel on lubricated steel	0.05	0.03
Teflon on steel	0.05–0.2	0.04
Teflon on Teflon	0.04	0.04
Ice on ice	0.1	0.02
Steel on ice	0.03	0.03
Wood on ice	0.05	0.05
Human synovial joints	0.016	0.015

Momentum

Momentum describes the amount of mass being displaced over time. Momentum is the product of mass and velocity, so it is a vector quantity. Momentum is often associated with the concept of inertia, since the momentum of an object is directly proportional to the force required to bring its motion to a halt. One of the most important aspects of momentum is that momentum is always conserved. Within a complete system, the total momentum remains the same. However, when only looking at a limited scope of a system, the momentum of an object in motion may be thought of as "lost," due to resistive forces like friction or air resistance. That "lost" momentum is being imparted to the ground or air, and as such the total momentum remains the same.

The following equation relates momentum to mass and velocity, where p is momentum, m is mass, and v is velocity:

$$p = mv$$

Most often, an object's velocity will change in a collision, but it is uncommon that there will be any change to its mass.

The concept of momentum is also fundamental to understanding interactions between objects, such as collisions. If a stationary object of a given mass is impacted by a moving object of some other mass, the stationary object will end up moving. The resulting velocity of the objects will depend on their masses, velocities, and whether the collision was elastic or inelastic. In an **elastic collision**, both kinetic energy and momentum are conserved. A completely elastic collision would result in the total momentum remaining the same. Because the total momentum before and after a collision is the same, the conservation of momentum can be mathematically represented by the following equation:

$$m_1 v_{1i} + m_2 v_{2i} = m_1 v_{1f} + m_2 v_{2f}$$

Similarly, the conservation of kinetic energy is represented as:

$$\frac{1}{2} m_1 v_{1i}^2 + \frac{1}{2} m_2 v_{2i}^2 = \frac{1}{2} m_1 v_{1f}^2 + \frac{1}{2} m_2 v_{2f}^2$$

Through algebraic rearrangement and factoring, the two equations can be rewritten to solve for the final velocities of two masses:

$$v_{1f} = \left(\frac{m_1 - m_2}{m_1 + m_2} \right) v_{1i} + \left(\frac{2 \times m_2}{m_1 + m_2} \right) v_{2i}$$

$$v_{2f} = \left(\frac{m_2 - m_1}{m_1 + m_2} \right) v_{2i} + \left(\frac{2 \times m_1}{m_1 + m_2} \right) v_{1i}$$

Note the change in sign and the subscripts of the variables.

One application of the elastic collision concept is the device known as **Newton's cradle**. A series of metal balls of equal mass are each hung at equal heights so they're in contact with each other when at rest. Raising one or more balls on one end and releasing them will allow them swing into the others, and the energy is then imparted to the stationary balls, causing them to swing up and back on the other side, while the previously moving balls are stationary.

In an **inelastic collision**, momentum in the system is still conserved, but kinetic energy is not. This loss of kinetic energy is due to it becoming physical deformation and heat. The result of an inelastic collision is the two objects moving in the same direction as each other. A completely inelastic collision means the objects are moving together in unison.

Projectile Motion

Projectile motion refers to the path an object takes when it moves through the air, solely under the effect of gravity. Due to the complexity in calculating projectile motion by hand when taking drag into account, it is often ignored. For heavier objects, the effect of air resistance becomes negligible as mass increases, because it has less effect on the object's momentum. Put simply, the path of a projectile in gravity will follow a downwards parabola.

Consider the following example. A football player is making a long throw and tosses the football into the air at an angle. Once he lets go, the ball is only being acted upon by gravity and air resistance (drag). When air resistance is ignored, solving projectile motion becomes simple: look at its motion in a plane. If the angle and magnitude of the initial velocity are known, the vertical and horizontal components can be looked at separately using trigonometry. An object thrown at a 30-degree angle at 10 meters per second will have an initial upwards velocity equal to the sine of 30 degrees times its initial velocity:

$$\sin(30°) \times 10\,\frac{\text{m}}{\text{s}} = \frac{1}{2} \times 10\,\frac{\text{m}}{\text{s}} = 5\,\frac{\text{m}}{\text{s}}$$

Thus, the ball will start out moving at 5 meters per second upwards, while being accelerated downwards by gravity at 9.81 meters per second squared. To find the change in height of an object, the following equation is used:

$$\Delta h = v_0 + \frac{1}{2}at^2$$

For a projectile with some initial upwards velocity, there will be two points where the change in height is zero: the start of its flight, and the point where it falls back to its initial height.

Circular Motion

Many mechanical systems involve rotating objects, such as the shafts and gear train in an engine.

An object in **circular motion** is constantly accelerating. The object is being accelerated, at every point, toward the center of the circle, so the direction of its acceleration is constantly changing. For an object moving at a constant speed, the magnitude of its acceleration remains constant. Without the acceleration towards the center, the object would continue to move in a straight line, in accordance with Newton's first law. Many celestial bodies exhibit orbits closely resembling circular motion.

Energy

Energy is the capacity to do work. In mechanical physics, energy is defined in terms of how much energy is needed to do a certain amount of work, which is measured in units called joules (J). **Work** is defined as the force applied over a distance in the direction of the force and is measured in the same units of energy. If force is applied to an object but it moves no distance in the same direction as that force, then no physical work has been done, despite effort being spent. The mechanical definition of work can be thought of as only effective force across a distance. One joule is equal to the energy transferred when one newton of force is applied to an object across a one-meter distance, thus the joule is also noted as a newton-meter (N · m).

Energy exists in several forms, but there are two main components to the total energy of an object—potential energy and kinetic energy. **Potential energy** is the energy stored in an object because of its physical composition or location, and **kinetic energy** is the energy of an object in motion. The total energy of a system is the sum of its potential energy and kinetic energy.

In basic physics, potential energy refers to the energy stored in an object as a function of its mass, its height, and gravity. The equation for potential energy is:

$$PE = mgh$$

Potential energy is represented by PE, the mass is m, acceleration due to gravity is g, and h is the height from which the object can fall, given some reference point (usually the ground). An object being held four feet in the air would be said to be at a height of four feet. The units for potential energy are a combination of the quantities that comprise it—mass, acceleration (gravity), and length (height). Since mass times acceleration is force, energy is just force times length. In SI units, this is a newton-meter, or N · m.

Kinetic energy exists in all things that are moving. The equation for kinetic energy is:

$$KE = \frac{1}{2}mv^2$$

The kinetic energy of an object is thus directly proportional to its mass, and to the square of its velocity. Because of this, small changes in velocity produce a larger change in the object's kinetic energy.

Angular Motion

The next step to mastering physics is to understand how linear concepts such as velocity, acceleration, force, and momentum apply to rotating objects. For objects that spin or move around an axis, their movement can be simplified using the concept of **angular motion**. Consider a baseball bat being swung. Intuitively, the force from the bat is greater towards its end, where it's moving faster. Although you might not know how fast the tip of the bat is moving, it's easier to measure how long it took to swing the bat through an angle. The rate the angle of the bat changes over time is its **angular velocity**:

$$\omega = \frac{\Delta\theta}{\Delta t}$$

Torque

Torque is the angular equivalent to force in rotational movement. Torque is equal to the perpendicular force applied to a lever, multiplied by the radius. Only the portion of the force that's perpendicular to the lever applies torque. The torque from a given force F applied at angle θ from the lever and at length r from the fulcrum fits the following equation:

$$\tau = F \times r \times \sin(\theta)$$

A useful application of these two properties is in calculating the power of a rotating object. Recall that power is a measure of work over time. The work performed in rotation can be calculated by the torque times the angle of rotation:

$$W = \tau\theta$$

Using this angular definition of work, power for angular motion becomes:

$$P = \tau\frac{\theta}{\Delta t} = \tau\omega$$

It's important to become familiar with algebraically rearranging, substituting, and combining physics equations to be able to solve for unknowns when approaching complex problems.

Machines

Machines are mankind's applications of physical principles to do work more efficiently. Although there are many varieties of complex machines for any task, there a few types of simple machines which are most used. Six of these include the lever, wedge, screw, wheel and axel, inclined plane, and pulley. For a simple machine that loses no energy to internal friction, it is termed an ideal simple machine.

Mechanical Advantage

The purpose of most machines is to make a task easier to perform by spreading out an input force over a distance. Consequently, the total work isn't reduced, but less force is needed in each time frame. The ability of a tool or machine to accomplish this is called its **mechanical advantage**, the amount that a machine amplifies input force. For a frictionless system, the mechanical advantage is simply equal to the force output through the machine over the force input, or $MA = \frac{F_{output}}{F_{input}}$. For real systems, the mechanical advantage includes a loss of energy in the output force. The ratio of the ideal mechanical advantage to the actual mechanical advantage of a machine is known as its **mechanical efficiency**, which represents how close the machine is to the ideal.

Levers

Levers are composed of a rigid body rotating in two dimensions on a point called a **fulcrum**. Levers are split into three classes, based on the relative positions of the fulcrum, load (or resistance), and the force (or effort) applied to it.

Assuming the lever is rigid, its mechanical advantage can be near to ideal. For an ideal lever, with a given input force F_E at distance a from the fulcrum and a load F_L at distance b from the fulcrum, the mechanical advantage is equal to the ratio of the effort and load forces:

$$F_E a = F_L b$$

A **class 1 lever** has the fulcrum between the load and the applied force. An example of a class 1 lever would be a seesaw, where there is a load on either side, pushing down the lever (the see-saw) in opposite directions.

A **class 2 lever** has the fulcrum located on one end, with the resistance in between the fulcrum and the applied force. Wheelbarrows are a common example of this. The wheelbarrow rests on its wheel at all times and is lifted by handles extending backwards, allowing the use of torque to raise the load more easily.

A **class 3 lever** applies force between the load and the fulcrum. A human forearm is an example of a class 3 lever. The arm bends at the elbow, while the weight of the arm acts as a mass at the end of the (relatively) rigid forearm. The forearm is flexed by a muscle and tendon attached just past the elbow, pulling the forearm towards the shoulder.

Wheel and Axles

The **wheel and axle** is effectively a special application of a lever. An **axle** is a rod that can be twisted, and a **wheel** is a circular shape on an end that force is applied through. The radius of the wheel, or the distance from the axle, is what allows for the amplification of force, through torque, the same as a lever. The weight of the axle and its resistance to turning is analogous to the load on a lever. Similar to the lever, the wheel and axle increases the distance the force is applied over to amplify the force. As a result, rotating a wheel by its outer rim amplifies the rotational force (torque) applied to the axle by a ratio of its wheel's radius to the radius of the axle. This reflects the mechanical advantage of the machine, which in equation form is:

$$MA = \frac{r_{wheel}}{r_{axle}}$$

220

Pulleys

The **pulley** is like the wheel and axle, but it adds another component: a rope, or some other sort of cable, is wound around it. The wheel of a pulley, called a **sheave**, is designed with a groove for a cable to sit in without slipping out. Since the sheave can rotate, pulling the cable spins the pulley rather than working entirely against friction to slide the rope. The benefit of the pulley is that for each attachment the pulley has to a fixed surface, the mechanical advantage increases, because that fraction of the load is being supported by its connection to the fixed surface. For example, when lifting an object directly, there is no additional mechanical advantage, but when load is suspended from a fixed point and a pulley, half of its weight is held by the fixed point. However, the cable must instead be pulled twice as far to perform the same work, because the load is held by two points—the fixed point and where it is being pulled.

The mechanical advantage of a pulley can be represented by the equation, where N is the number of fixed points holding the load:

$$MA = N$$

Inclined Planes

An **inclined plane** is any sloped surface. Typically, they are used to displace an object in one axis perpendicular to another. The best example is a ramp, which raises or lowers over a lateral distance. The inclined plane is one of the best examples of mechanical advantage, because the ideal mechanical advantage is equal to the inverse of the slope. In other words, the mechanical advantage without friction is simply length of the sloped surface divided by its height. Take note that this is not the horizontal length of the inclined plane, but the length of the slope. For an inclined plane that's nine feet long and three feet tall, the mechanical advantage would be:

$$\text{MA} = \frac{l}{h} = \frac{9\text{ ft}}{3\text{ ft}} = 3$$

Wedges

Wedges are effectively mobile inclined planes. A common example of a wedge is an axe head. The axe head is wide near the shaft, and tapers down to its edge. The mechanical advantage for a wedge is the same as that of an inclined plane—length divided by its width or height (in the direction it's pushing against). One difference is that the wedge is typically thought of as being driven into something (such as an axe) or between two things (such as a forked path on a conveyor belt), rather than a fixed surface.

Screws

Screws are another application of the inclined plane. In essence, the tip of a screw behaves like a conical wedge, pushing wood (or another material) to the sides of the screw as it is pushed forward, while the threading on the screw acts like an inclined plane by pushing the wood backwards (or pulling the screw inwards) as the screw threading moves into the wood. The mechanical advantage of an ideal screw is the circumference of the screw shaft divided by the thread lead:

$$\frac{2\pi r}{l}$$

In the previous equation, r represents the radius of the screw's shaft, while l is the screw's **lead**, the length it takes for a thread to wrap once around the shaft. A screw with multiple threads will have a different lead and pitch. The **pitch** of a screw is instead the distance from one thread to the next. Screws with multiple threads will have differing lead and pitch—a screw with two symmetrical threads will have a pitch half the length of its lead.

Practice Quiz

1. What is mass?
    a. Mass is an object's resistance to a change in motion.
    b. Mass measures the entire quantity of an object's matter.
    c. Mass is the amount of force that is exerted onto an object due to a gravitational pull.
    d. Mass will change depending on its environment.

2. If a vehicle is moving with an acceleration of 4 m/s^2 with an applied force of 5,000 N, what is the mass of the vehicle, assuming that friction is not a factor?
    a. 1,200 kg
    b. 1,250 kg
    c. 1,350 kg
    d. 1,400 kg

3. What is the name for a force that resists movement due to the rubbing between two surfaces?
    a. The coefficient of friction
    b. The normal force
    c. Kinetic friction
    d. Static friction

4. Find the weight of an object with a mass of 10 kg object given that the normal force holding the object is equal to the weight.
    a. 98.0 N
    b. 0.98 N
    c. 0.98 m/s^2
    d. 98 m/s^2

5. Suppose that a person applies a 60 N continuous force, in a direction parallel to the floor, towards a box that has a weight of 100 N. If the coefficient of kinetic friction is $\mu_k = 0.25$ and the coefficient of static friction is $\mu_s = 0.50$, what will happen to the box?
    a. The box will start to accelerate.
    b. The box will move briefly but then stop since the kinetic friction is greater than the applied force.
    c. The box will move forward but then bounce back due to friction.
    d. The box will stay at rest since the applied force is less than the static friction.

See answers on next page.

Answer Explanations

1. B: Choice *B* is correct and describes an intrinsic property of mass. Choice *A* is incorrect and is the definition for inertia. Choice *C* is incorrect and is the definition of weight. Choice *D* is incorrect since mass remains the same regardless of the environment. For instance, an astronaut's mass is the same on Earth and the Moon. The weight of an astronaut is greater on Earth than on the Moon.

2. B: Choice *B* is correct. Newton's second law can be rearranged to calculate the mass of the vehicle.

$$F = ma$$

F: force in Newtons (N)

m: mass in kilograms (kg)

a: acceleration in meters per second square m/s^2

First, rearrange the equation and then solve for the mass.

$$m = \frac{F}{a} = \frac{5,000 \text{ N}}{4 \,^{\text{m}}/_{\text{s}^2}} = 1,250 \text{ kg}$$

3. C: The correct answer is kinetic friction, Choice *C*. Kinetic friction is a responsive force that will resist movement when two surfaces rub against one another. For example, there is kinetic friction when an object moves across a surface. Kinetic friction acts in a direction opposite to the movement of the object. Choice *A*, the coefficient of friction, which is incorrect, represents the amount of friction between two material surfaces that move or slide past one another. Choice *B*, the normal force, which is incorrect, is the opposite and equal force that a surface will exert when pressed by an object. Choice *D*, static friction, is the friction present between two surfaces or objects that prevents movement; it is the friction that occurs when trying to move a stationary object without initiating movement between the two surfaces.

4. A: Choice *A* is the correct answer. The weight of any object is given by the following equation:

$$W = F_N = mass \times gravity \ of \ Earth$$

The normal force (F_N) is equal to the weight only when the object is lying flat on a surface. The mass is equal to 10 kg, and the average standard gravity or gravity of Earth is 9.8 m/s^2. The weight of the object is:

$$W = F_N = 10 \text{ kg} \times 9.8 \frac{\text{m}}{\text{s}^2} = 98 \text{ N}$$

The correct units for weight are Newtons (N) since kilograms multiplied by meters per second squared are equivalent to a Newton.

5. A: Choice *A* is correct. The normal force (F_N) or weight of an object that is lying flat on a surface has already been given.

$$W = F_N = 100 \text{ N}$$

Use the gravitational force to find the mass of the box ($mass = \frac{force}{gravity}$). The mass of the box is:

$$m = \frac{100 \text{ N}}{9.8 \frac{\text{m}}{\text{s}^2}} \approx 10 \text{ kg}$$

Next, calculate the kinetic and static frictional forces using the following equations:

$$F_k = \mu_k F_N = (0.25)(100 \text{ N}) = 25.0 \text{ N}$$

$$F_s = \mu_s F_N = (0.50)(100 \text{ N}) = 50.0 \text{ N}$$

Since the applied force of 60 N is greater than the static frictional forces, the object will move and accelerate, which makes Choice A the correct answer. Total net force, excluding the static force, is simply the difference between the applied force and the kinetic force:

$$F_{net} = F_{applied} - F_{kinetic} = 60 \text{ N} - 25 \text{ N} = 35 \text{ N}$$

The acceleration is positive:

$$F_{net} = ma \text{ or } 35 \text{ N} = 10 \text{ kg} \times a$$

$$a = \frac{35 \text{ N}}{10 \text{ kg}} = 3.5 \text{ m/s}^2$$

Choice B would be partly correct if the applied force of 60 N was only applied briefly. Choice C is not correct since the applied force is only in one direction. The kinetic force will not cause the box to move backward and will only cause the box to slow down or decelerate if the applied force is not continuous. Choice D is incorrect since the applied force is greater than the static frictional force.

Assembling Objects

The computer adaptive testing Armed Services Vocational Aptitude Battery (ASVAB) may include a minimum of fifteen scored questions on assembling objects (AO); these have subtest time limit of seventeen minutes. However, fifteen possible try-out questions extend the subtest time limit to thirty-six minutes. The purpose of these AO questions is to determine how individual objects look when assembled. These problems test a person's visual-spatial ability and how well they can put irregularly shaped pieces together. In addition to the ASVAB, AO questions are often included in tests in law enforcement, mechanics, and engineering. Spatial relationship skills can be improved by practicing and adapting specific learning strategies.

Assembling Objects (AO) consists of two or more objects such as triangles, rectangles, and irregularly shaped objects. These objects may include lines and shapes, each with a labeled point and letter. An assembling object problem will contain two or more unique objects.

There are two types of test problems in the AO section: jigsaw puzzle and connector type problems. Jigsaw puzzle problems test a person's ability to choose how an object looks when all its individual parts are combined to form a larger shape. Connector-type problems test a person's ability to connect two parts with a line at a specified point, typically designated by a letter. For each problem type, there is a total of five connected boxes. The first box on the left contains drawings of individual components that need to be connected. The next four boxes, which may be labeled A through D, contain several possible shapes. Each box contains a possible shape that may result when each individual component, shown in the first box, is correctly assembled. The figure below is a jigsaw puzzle problem and shows unassembled objects in the leftmost box.

Figure 1. Jigsaw puzzle problem

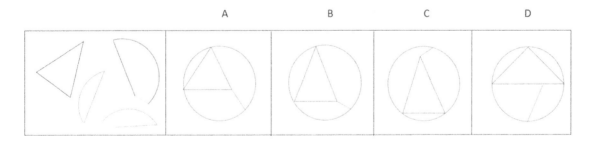

Figure 1 shows the unassembled object in the leftmost box. This box of components is referred to as an instructions box. Choices A through D represent various possible assembled structures. One of the choices is correct, while the other three are incorrect. The correct choice will contain the correct shapes that fit together to make a complete circle. One effective strategy for identifying the correct choice is to look for a common point of reference. For example, there is one triangle in the instructions box. Review each possible answer choice to visually identify the correct shape. For example, Choice A can be eliminated since it is not an isosceles triangle. Choice D is incorrect since the isosceles triangle has a smaller height. Since Choice C contains a similarly shaped triangle, other shapes must be looked at. For example, Shape 2 in Choice C is smaller; Shapes 3 and 4 are also different at the edges. Therefore, through a process of elimination, Choice B is the best answer.

	A	B	C	D

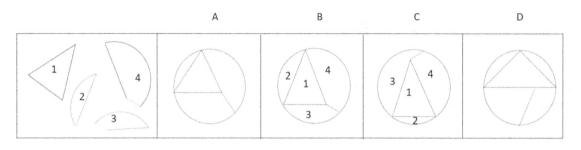

Practice Quiz

1.

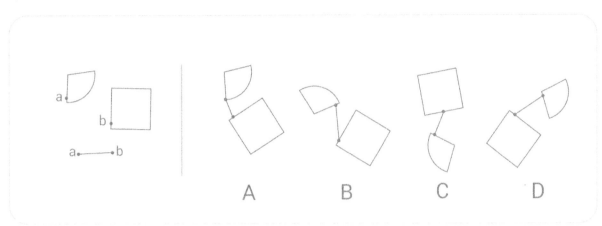

2.

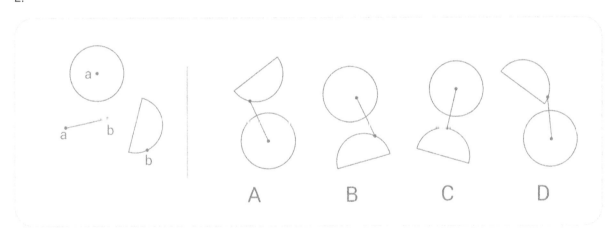

3.

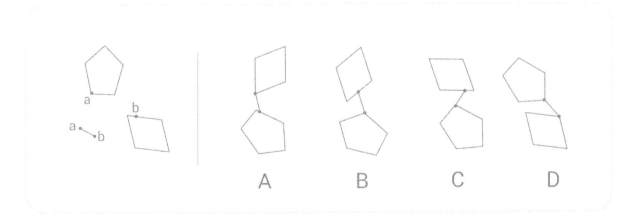

4.

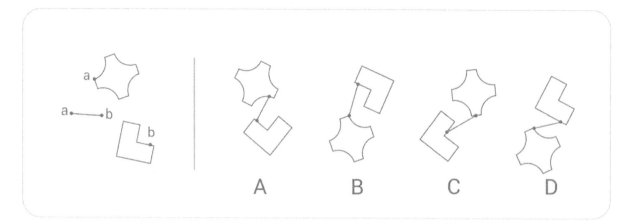

5.

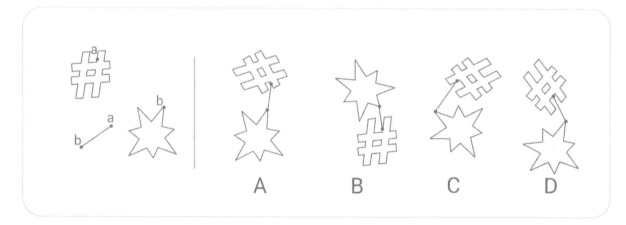

Answer Explanations

1. B

2. A

3. C

4. B

5. C

See answers on next page.

Practice Test #1

General Science

1. When clothes are being spun in a washing machine, what type of force causes the water to be expelled through the small holes in the barrel of the washer?
    a. Centripetal
    b. Friction
    c. Angular momentum
    d. Centrifugal

2. Under which situation does an object have a negative acceleration?
    a. It is increasing velocity in a positive direction.
    b. It is at a complete stop.
    c. It is increasing velocity in a negative direction.
    d. It is moving at a constant velocity.

3. What part of the respiratory system is responsible for regulating the temperature and humidity of the air that comes into the body?
    a. Larynx
    b. Lungs
    c. Trachea
    d. Sinuses

4. What type of reactions involve the breaking and re-forming of bonds between atoms?
    a. Chemical
    b. Physical
    c. Isotonic
    d. Electron

5. Which organ is responsible for gas exchange between air and circulating blood?
    a. Nose
    b. Larynx
    c. Lungs
    d. Stomach

6. Which molecule is the simplest form of sugar?
    a. Monosaccharide
    b. Fatty acid
    c. Polysaccharide
    d. Amino acid

7. Which of the following is a distinct characteristic of a gas?
    a. It has a fixed shape.
    b. It is easy to compress.
    c. It has an exact volume.
    d. The molecules are packed together tightly.

8. Which component of blood helps to fight off diseases?
   a. Red blood cells
   b. White blood cells
   c. Plasma
   d. Platelets

9. Which system consists of a group of organs that work together to transform food and liquids into fuel for the body?
   a. Respiratory system
   b. Immune system
   c. Genitourinary system
   d. Gastrointestinal system

10. During which step performed by the gastrointestinal system do chemicals and enzymes break down complex food molecules into smaller molecules?
   a. Digestion
   b. Absorption
   c. Compaction
   d. Ingestion

11. Which of the following is an example of a physical property of substances?
   a. Odor
   b. Reactivity
   c. Flammability
   d. Toxicity

12. Why is cellular reproduction essential for an organism?
   a. It allows for the replacement of dying and damaged cells.
   b. It marks the end of the organism's life.
   c. It marks the end of growth for an organism.
   d. It makes changes to the organism's DNA.

13. When is a reactant characterized as an oxidative agent in a redox reaction?
   a. When it gains a hydrogen
   b. When it loses an oxygen
   c. When it gains one or more electrons
   d. When it gains a carbon

14. What is the sum of the oxidation numbers of all the atoms in a neutral compound?
   a. 1
   b. −1
   c. 2
   d. 0

15. What type of light is harmful to human skin?
   a. Ultraviolet (UV) light
   b. The whole visible spectrum
   c. Radio waves
   d. Blue light

Arithmetic Reasoning

1. In a neighborhood, 15 out of 80 of the households have children under the age of 18. What percentage of the households have children under 18?
   a. 0.1875%
   b. 18.75%
   c. 1.875%
   d. 15%

2. Gina took an algebra test last Friday. There were 35 questions, and she answered 60% of them correctly. How many correct answers did she have?
   a. 35
   b. 20
   c. 21
   d. 25

3. Paul took a written driving test, and he got 12 of the questions correct. If he answered 75% of the total questions correctly, how many questions were on the test?
   a. 25
   b. 16
   c. 20
   d. 18

4. If a car is purchased for $15,395 with a 7.25% sales tax, how much is the total price?
   a. $15,395.07
   b. $16,511.14
   c. $16,411.13
   d. $15,402

5. A car manufacturer usually makes 15,412 SUVs, 25,815 station wagons, 50,412 sedans, 8,123 trucks, and 18,312 hybrids a month. About how many cars are manufactured each month?
   a. 120,000
   b. 200,000
   c. 300,000
   d. 12,000

6. A family goes to the grocery store every week and spends $105. About how much does the family spend annually on groceries?
   a. $10,000
   b. $50,000
   c. $500
   d. $5,000

7. Bindee is having a barbeque on Sunday and needs 12 packets of ketchup for every 5 guests. If 60 guests are coming, how many packets of ketchup should she buy?
   a. 100
   b. 12
   c. 144
   d. 60

232

8. A grocery store sold 48 bags of apples in one day. If 9 of the bags contained Granny Smith apples and the rest contained Red Delicious apples, what is the ratio of bags of Granny Smith to bags of Red Delicious that were sold?
   a. 48:9
   b. 39:9
   c. 9:48
   d. 9:39

9. If Oscar's bank account totaled $4,000 in March and $4,900 in June, what was the rate of change in his bank account total over those three months?
   a. $900 a month
   b. $300 a month
   c. $4,900 a month
   d. $100 a month

10. The percentage of smokers above the age of 18 in 2000 was 23.2%. The percentage of smokers above the age of 18 in 2015 was 15.1%. Find the average rate of change in the percent of smokers above the age of 18 from 2000 to 2015.
   a. −0.54%
   b. −54%
   c. −5.4%
   d.   15%

11. To estimate deer population in a forest, biologists obtained a sample of deer in that forest and tagged each one of them. The sample had 300 deer in total. They returned a week later, harmlessly captured 400 deer and found that 5 were tagged. Use this information to estimate how many total deer were in the forest.
   a. 24,000 deer
   b. 30,000 deer
   c. 40,000 deer
   d. 80,000 deer

12. The number of members of the House of Representatives varies directly with the total population in a state. If the state of New York has 20,000,000 residents and has 27 total representatives, how many should Ohio have with a population of 11,600,000?
   a. 10
   b. 16
   c. 11
   d. 5

13. At a mall, 45% of the stores are clothing stores. If there are 60 stores in the mall, how many are not clothing stores?
   a. 27
   b. 33
   c. 30
   d. 23

14. Steve rented a delivery truck to deliver his flower arrangements. He rented the truck at 9 a.m. and returned it at 6 p.m. He paid a total of $140.85 for the truck. What was the rental cost per hour?
   a. $140.85
   b. $15.65
   c. $14.09
   d. $17.61

15. Katy works at a shoe store. She makes $15 an hour and works 20 hours each week. Also, she earns a 6% commission on all her sales. If Katy sells $2,400 worth of shoes in a single week, how much are her total earnings for the week?
   a. $300
   b. $144
   c. $444
   d. $2,100

Word Knowledge

1. *Wary* most nearly means
   a.   Religious
   b.   Adventurous
   c.   Tired
   d.   Cautious

2. *Proximity* most nearly means
   a.   Estimate
   b.   Delicate
   c.   Precarious
   d.   Closeness

3. *Parsimonious* most nearly means
   a.   Lavish
   b.   Harmonious
   c.   Miserly
   d.   Careless

4. *Propriety* most nearly means
   a.   Ownership
   b.   Appropriateness
   c.   Patented
   d.   Abstinence

5. *Boon* most nearly means
   a.   Cacophony
   b.   Hopeful
   c.   Benefit
   d.   Squall

6. *Strife* most nearly means
   a. Plague
   b. Industrial
   c. Picketing
   d. Conflict

7. *Qualm* most nearly means
   a. Calm
   b. Uneasiness
   c. Assertion
   d. Pacify

8. *Valor* most nearly means
   a. Rare
   b. Coveted
   c. Leadership
   d. Bravery

9. *Zeal* most nearly means
   a. Craziness
   b. Resistance
   c. Fervor
   d. Opposition

10. *Extol* most nearly means
    a. Glorify
    b. Demonize
    c. Chide
    d. Admonish

11. *Reproach* most nearly means
    a. Locate
    b. Chide
    c. Concede
    d. Orate

12. *Milieu* most nearly means
    a. Bacterial
    b. Damp
    c. Ancient
    d. Environment

13. *Guile* most nearly means
    a. Masculine
    b. Stubborn
    c. Naïve
    d. Deception

14. *Assent* most nearly means
    a. Acquiesce
    b. Climb
    c. Assert
    d. Demand

15. *Dearth* most nearly means
    a. Grounded
    b. Scarcity
    c. Lethal
    d. Risky

Paragraph Comprehension

Directions for Questions 1–15: After reading the passage, choose the best answer to the question based on what is stated in the passage.

1. Rehabilitation, rather than punitive justice, is becoming much more popular in prisons around the world. Prisons in America, especially, where the recidivism rate is 67 percent, would benefit from mimicking prison tactics in Norway, which has a recidivism rate of only 20 percent. In Norway, the idea is that a rehabilitated prisoner is much less likely to offend than one harshly punished. Rehabilitation includes proper treatment for substance abuse, psychotherapy, healthcare and dental care, and education programs.

Which of the following best captures the author's purpose?
    a. To show the audience one of the effects of criminal rehabilitation by comparison
    b. To persuade the audience to donate to American prisons for education programs
    c. To convince the audience of the harsh conditions of American prisons
    d. To inform the audience of the incredibly lax system of Norwegian prisons

2. What a lark! What a plunge! For so it had always seemed to her, when, with a little squeak of the hinges, which she could hear now, she had burst open the French windows and plunged at Bourton into the open air. How fresh, how calm, stiller than this of course, the air was in the early morning; like the flap of a wave; the kiss of a wave; chill and sharp and yet (for a girl of eighteen as she then was) solemn, feeling as she did, standing there at the open window, that something awful was about to happen; looking at the flowers, at the trees with the smoke winding off them and the rooks rising, falling; standing and looking until Peter Walsh said, "Musing among the vegetables?"
                     From <u>Mrs. Dalloway</u> by Virginia Woolf

What was the narrator feeling right before Peter Walsh's voice distracted her?
    a. A spark of excitement for the morning
    b. Anger at the larks
    c. A sense of foreboding
    d. Confusion at the weather

3. According to the plan of the convention, all judges who may be appointed by the United States are to hold their offices *during good behavior*, which is conformable to the most approved of the State constitutions and among the rest, to that of this State. Its propriety having been drawn into question by the adversaries of that plan, is no light symptom of the rage for objection, which disorders their imaginations and judgments. The standard of good behavior for the continuance in office of the judicial magistracy, is certainly one of the most valuable of the modern improvements in the practice of government. In a monarchy it is an excellent barrier to the despotism of the prince; in a republic it is a no less excellent barrier to the encroachments and oppressions of the representative body. And it is the best expedient which can be devised in any government, to secure a steady, upright, and impartial administration of the laws.

From <u>The Federalist No. 78</u> by Alexander Hamilton

What is Hamilton's point in this excerpt?
 a. To show the audience that despotism within a monarchy is no longer the standard practice in the states
 b. To convince the audience that judges holding their positions based on good behavior is a practical way to avoid corruption
 c. To persuade the audience that having good behavior should be the primary characteristic of a person in a government body and that their voting habits should reflect this
 d. To convey the position that judges who serve for a lifetime will not be perfect and, therefore, we must forgive them for their bad behavior when it arises

4. There was a man named Webster who lived in a town of twenty-five thousand people in the state of Wisconsin. He had a wife named Mary and a daughter named Jane and he was himself a fairly prosperous manufacturer of washing machines. When the thing happened of which I am about to write, he was thirty-seven or thirty-eight years old and his one child, the daughter, was seventeen. Of the details of his life up to the time a certain revolution happened within him it will be unnecessary to speak.

From <u>Many Marriages</u> by Sherwood Anderson

What does the author mean by the following sentence?

"Of the details of his life up to the time a certain revolution happened within him it will be unnecessary to speak."
 a. The details of his external life don't matter; only the details of his internal life matter.
 b. Whatever happened in his life before he had a certain internal change is irrelevant.
 c. He had a traumatic experience earlier in his life which rendered it impossible for him to speak.
 d. Before the revolution, he was a lighthearted man who always wished to speak to others no matter who they were.

5. The Prince was untiring in planning improvements, and in 1856 the Queen wrote: "Every year my heart becomes more fixed in this dear Paradise, and so much more so now, that *all* has become my dearest Albert's *own* creation, own work, own building, own laying out as at Osborne; and his great taste, and the impress of his dear hand, have been stamped everywhere. He was very busy today, settling and arranging many things for next year."

From the biography <u>Queen Victoria</u> by E. Gordon Browne, M.A.

What does the word *impress* mean in the third paragraph?
   a. To affect strongly in feeling
   b. To urge something to be done
   c. A certain characteristic or quality imposed upon something else
   d. To press a thing onto something else

6. Booth who was even more fashionably and richly dressed than usual, walked thence around to the front of the theater, and went in. Ascending to the dress circle, he stood for a little time gazing around upon the audience and occasionally upon the stage in his usual graceful manner. He was subsequently observed by Mr. Ford, the proprietor of the theater, to be slowly elbowing his way through the crowd that packed the rear of the dress circle toward the right side, at the extremity of which was the box where Mr. and Mrs. Lincoln and their companions were seated. Mr. Ford casually noticed this as a slightly extraordinary symptom of interest on the part of an actor so familiar with the routine of the theater and the play.
   From <u>The Life, Crime, and Capture of John Wilkes Booth</u> by George Alfred Townsend

What does the author mean by the last two sentences?
   a. Mr. Ford was suspicious of Booth and assumed he was making his way to Mr. Lincoln's box.
   b. Mr. Ford assumed Booth's movement throughout the theater was due to being familiar with the theater.
   c. Mr. Ford thought that Booth was making his way to the theater lounge to find his companions.
   d. Mr. Ford thought that Booth was elbowing his way to the dressing room to get ready for the play.

7. When we study more carefully the effect upon the milk of the different species of bacteria found in the dairy, we find that there is a great variety of changes which they produce when they are allowed to grow in milk. The dairyman experiences many troubles with his milk. It sometimes curdles without becoming acid. Sometimes it becomes bitter, or acquires an unpleasant "tainted" taste, or, again, a "soapy" taste.
   From <u>The Story of Germ Life</u> by Herbert William Conn

What is the tone of this passage?
   a. Excitement
   b. Anger
   c. Neutral
   d. Sorrowful

8. It is perhaps true of many of the growing cities of the West, that they do not offer the same social advantages as the older cities of the East. But this is principally the case as to what may be called boom cities, where the larger part of the population is of that floating class which follows in the line of temporary growth for the purposes of speculation, and in no sense applies to those centers of trade whose prosperity is based on the solid foundation of legitimate business. As the metropolis of a vast section of country, having broad agricultural valleys filled with improved farms, surrounded by mountains rich in mineral wealth, and boundless forests of as fine timber as the world produces, the cause of Portland's growth and prosperity is the trade which it has as the center of collection and distribution of this great wealth of natural resources, and it has attracted, not the boomer and speculator, who find their profits in the wild excitement of the boom, but the merchant, manufacturer, and investor, who seek the surer if slower channels of legitimate business and investment.

From <u>Oregon, Washington, and Alaska. Sights and Scenes for the Tourist</u>, written by E.L. Lomax in 1890

What is a characteristic of a "boom city," as indicated by the passage?
   a. It is built on solid business foundation of mineral wealth and farming
   b. It is an area of land on the west coast that quickly becomes populated by residents from the east coast
   c. Due to the hot weather and dry climate, it catches fire frequently, resulting in a devastating population drop
   d. Its population is made up of people who seek quick fortunes rather than building a solid business foundation

9. The other of the minor deities at Nemi was Virbius. Legend had it that Virbius was the young Greek hero Hippolytus, chaste and fair, who learned the art of venery from the centaur Chiron, and spent all his days in the greenwood chasing wild beasts with the virgin huntress Artemis (the Greek counterpart of Diana) for his only comrade.

From <u>The Golden Bough</u> by Sir James George Frazer

Based on a prior knowledge of literature, the reader can infer this passage is taken from which of the following?
   a. A eulogy
   b. A myth
   c. A historical document
   d. A technical document

10. When I wrote the following passages, or rather the bulk of them, I lived alone, in the woods, a mile from any neighbor, in a house which I had built myself on the shore of Walden Pond, in Concord, Massachusetts, and earned my living by the labor of my hands only. I lived there two years and two months. At present I am a sojourner in civilized life again.

From <u>Walden</u> by Henry David Thoreau

What does the word *sojourner* most likely mean at the end of the passage?
   a. Illegal immigrant
   b. Temporary resident
   c. Lifetime partner
   d. Farm crop

Mathematics Knowledge

1. Which expression is equivalent to $\sqrt[4]{x^6} - \frac{x}{x^3} + x - 2$?

    a. $x^{\frac{3}{2}} - x^2 + x - 2$

    b. $x^{\frac{2}{3}} - x^{-2} + x - 2$

    c. $x^{\frac{3}{2}} - \frac{1}{x^2} + x - 2$

    d. $x^{\frac{2}{3}} - \frac{1}{x^2} + x - 2$

2. How many possible positive zeros does the polynomial function $f(x) = x^4 - 3x^3 + 2x + x - 3$ have?
    a. 4
    b. 3
    c. 2
    d. 1

3. What is the solution to the following linear inequality?

$$7 - \frac{4}{5}x < \frac{3}{5}$$

    a. $(-\infty, 8)$
    b. $(8, \infty)$
    c. $[8, \infty)$
    d. $(-\infty, 8]$

4. Triple the difference of five and a number is equal to the sum of that number and 5. What is the number?
    a. 5
    b. 2
    c. 5.5
    d. 2.5

5. What is an equivalent form of the rational expression $\sqrt{200x^6y^7z^2}$? Assume all variables represent positive real numbers.
    a. $20x^3y^3z\sqrt{y}$
    b. $10x^3y^3z\sqrt{2y}$
    c. $10xyz\sqrt{2xyz}$
    d. $10x^2y^3z\sqrt{2z}$

6. What is the simplified form of $(4y^3)^4(3y^7)^2$?
    a. $12y^{26}$
    b. $2{,}304y^{16}$
    c. $12y^{14}$
    d. $2{,}304y^{26}$

240

7. Which equation correctly shows how to find the surface area of a cylinder?

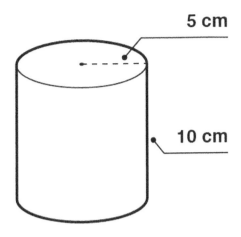

   a. $SA = 2\pi \times 5 \times 10 + 2\pi \times 5^2$
   b. $SA = 5 \times 2\pi \times 5$
   c. $SA = 2\pi \times 5^2$
   d. $SA = 2\pi \times 10 + \pi \times 5^2$

8. What is the area of the shaded region?

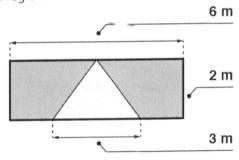

   a. $9 \, \text{m}^2$
   b. $12 \, \text{m}^2$
   c. $6 \, \text{m}^2$
   d. $8 \, \text{m}^2$

9. What is $\frac{12}{60}$ converted to a percentage?
   a. 0.20
   b. 20%
   c. 25%
   d. 12%

10. Which of the following is the correct decimal form of the fraction $\frac{14}{33}$ rounded to the nearest hundredth place?

    a. 0.44

    b. 0.42

    c. 0.424

    d. 0.140

11. Which of the following represents the correct sum of $\frac{14}{15}$ and $\frac{2}{5}$, in lowest possible terms?

    a. $\frac{20}{15}$

    b. $\frac{4}{3}$

    c. $\frac{16}{20}$

    d. $\frac{4}{5}$

12. What is the product of $\frac{5}{14}$ and $\frac{7}{20}$, in lowest possible terms?

    a. $\frac{1}{8}$

    b. $\frac{35}{280}$

    c. $\frac{12}{34}$

    d. $\frac{1}{2}$

13. What is the result of dividing 24 by $\frac{8}{5}$, in lowest possible terms?

    a. $\frac{5}{3}$

    b. $\frac{3}{5}$

    c. $\frac{120}{8}$

    d. 15

14. Subtract $\frac{5}{14}$ from $\frac{5}{24}$. Which of the following is the correct result?

    a. $\frac{25}{168}$

    b. 0

    c. $-\frac{25}{168}$

    d. $\frac{1}{10}$

15. Which of the following is a correct mathematical statement?

a. $\frac{1}{3} < -\frac{4}{3}$

b. $-\frac{1}{3} > \frac{4}{3}$

c. $\frac{1}{3} > -\frac{4}{3}$

d. $-\frac{1}{3} \geq \frac{4}{3}$

Electronics Information

1. In a series circuit, which is true of the total resistance?
    a. It is equal to the product of all resistances
    b. It is equal to the sum of all resistances
    c. It is greater than the sum of all resistances
    d. It is less than the sum of all resistances

2. Which of the following equations correctly represents the relationship of current, voltage, resistance, and power?
    a. $V = IP$
    b. $P = IR$
    c. $PR = V^?$
    d. $P = I^2V$

3. In the following circuit, which point has the LEAST current?

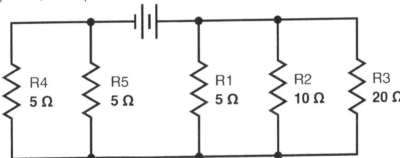

    a. R_1
    b. R_2
    c. R_3
    d. R_4

4. If resistance remains constant and power in a circuit decreases, what must happen to the current and voltage?
    a. Current and voltage both decrease
    b. Current increases while voltage remains constant
    c. Voltage increases while current remains constant
    d. Current and voltage both increase

5. What is the electric potential difference between points A and B?

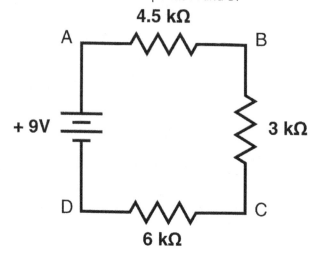

    a.   3 mV
    b.   4.5 mV
    c.   3V
    d.   4.5 V

6. What is the voltage of the power source in the following circuit?

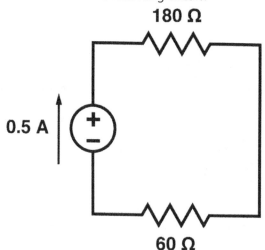

    a.   60 V
    b.   120 V
    c.   180 V
    d.   240 V

7. How does charge move through a simple circuit and in which direction is electron flow?
    a.   Charge is moving protons in a circuit from the positive terminal to the negative terminal.
    b.   Charge is moving electrons in a circuit from the positive terminal to the negative terminal.
    c.   Charge is moving protons in a circuit from the negative terminal to the positive terminal.
    d.   Charge is moving electrons in a circuit from the negative terminal to the positive terminal.

8. Household electricity is commonly measured using the kilowatt-hour (kWh), a unit of energy. Which of the following quantities is equivalent to the kilowatt-hour?
   a. voltage × charge
   b. work ÷ time
   c. power ÷ time
   d. energy × current

9. What is the voltage drop across R_2 in the following circuit?

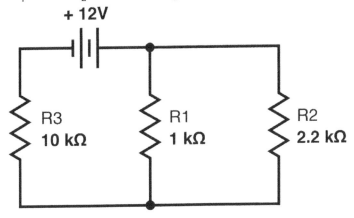

   a. 0.68 V
   b. 0.77 V
   c. 1.12 V
   d. 2.46 V

10. Which unit is a measure of energy?
    a. Joules
    b. Watts
    c. Coulombs
    d. Amperes

11. Which statement does NOT contribute to why current stops in an open (broken) circuit?
    a. Current stops because the electrons have nowhere else to move.
    b. Current stops because there is no charge difference.
    c. Current stops because there is no resistance.
    d. Current stops because the power source is disconnected.

12. According to Kirchhoff's current law, what is the unknown current through I in the following circuit segment?

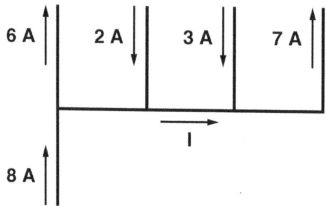

   a.   5 A
   b.   13 A
   c.   8 A
   d.   4 A

13. What must the resistance of R_x be if the voltmeter in the following diagram reads 0? Round to the nearest tenth of an ohm.

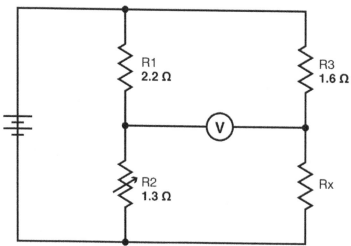

   a.   0.7 Ω
   b.   1.0 Ω
   c.   1.2 Ω
   d.   1.9 Ω

14. Which property is inversely proportional to conductance?
   a.   Resistance
   b.   Impedance
   c.   Reactance
   d.   Capacitance

15. A certain capacitor has a dielectric constant of 2.3 and a separation of 5.0 millimeters. Assuming the capacitor's plates are identical, what must the area of one of its plates be if its capacitance is 5.3 picofarads?

a. 1.30 mm²
b. 26.09 mm²
c. 11.52 mm²
d. 2.44 mm²

Auto Information

1. At which point in the engine's cycle does the spark plug ignite fuel?

a. After the intake valve opens
b. When the piston begins to compress the fuel-air mixture
c. Once the piston fully extends
d. After the exhaust valve opens

2. Which tool is best suited for dislodging a jammed bolt?

a. Needle nose pliers
b. Dead blow hammer
c. Vise grip
d. Ball-peen hammer

3. Which part of the ignition system increases voltage?

a. The battery
b. The distributor
c. The ignition control module
d. The ignition coil

4. Which of the following is NOT a part of a hydraulic brake system?

a. Calipers
b. Drum
c. Drive shaft
d. Master cylinder

5. Which of the following emissions is NOT reduced by the vehicle's catalytic converter?

a. Carbon monoxide
b. Nitrogen dioxide
c. Hydrocarbons
d. Nitrogen oxide

6. Which of the following is part of an ancillary circuit in the vehicle's electrical system?

a. Mass airflow sensors
b. Engine oil pump
c. Oxygen sensors
d. Starter engine

7. Which of the following is a reliable lift point on a vehicle?
   a. The tires
   b. The engine
   c. The drive shaft
   d. The differential

8. Which diagnostic trouble code would alert you to a general issue with the engine's ignition?
   a. P0300
   b. U1803
   c. B1301
   d. C0103

9. Which component is NOT part of a vehicle's lubrication system?
   a. Oil sump
   b. Transmission fluid
   c. Oil pump
   d. Oil feed lines

10. Which part of the engine converts linear motion into rotational motion?
   a. Differential
   b. Camshaft
   c. Crankshaft
   d. Gearbox

Shop Information

1. Which tool works best for cutting a piece of metal pipe that is 1 inch thick in diameter?
   a. Hacksaw
   b. Crosscut saw
   c. Jigsaw
   d. Chainsaw

2. Which piece of equipment helps reduce the risk of facial injury due to flying debris?
   a. Gloves
   b. Face shield
   c. Apron
   d. Face mask

3. Which tool, when used with a hammer, allows a worker to cut and shape different materials, including wood and stone?
   a. Nail
   b. Wedge
   c. Chisel
   d. Handsaw

4. A hammer with a semi-hollow head that is filled with sand is designed to produce minimum recoil. What is this type of hammer called?
   a. Soft blow hammer
   b. Dead blow hammer
   c. Soft strike hammer
   d. Dead strike hammer

5. What is the purpose of an exhaust hood?
   a. To capture and remove harmful inhalants from the workshop area
   b. To cover the exhaust fan
   c. To minimize noise from the ventilation system
   d. To collect the contaminant particles in the ventilation system

6. Which type of holding tool, normally attached to a table or work bench, is displayed in the image?

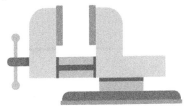

   a. Pliers
   b. Clamp
   c. Wrench
   d. Vice

7. Which tool would a carpenter use to shave away thin strips from the edge of a wooden board?
   a. Drill
   b. Bandsaw
   c. Planer
   d. Level

8. Which of the following tools would be used to cut wood along (or parallel to) the grain?
   a. Fret saw
   b. Rip saw
   c. Crosscut saw
   d. Hacksaw

9. Which tool is NOT adjustable to different sized tasks?
   a. Open-end wrench
   b. Crescent wrench
   c. Monkey wrench
   d. Pliers

10. Which task would a wooden router be used for?
    a. Cutting a wooden board in half
    b. Making holes in wooden boards
    c. Engraving designs into wood
    d. Making the edges of a board smooth

Mechanical Comprehension

1. For the following block on a slope, what formula represents the normal force acting upon the block, given the block is not moving and the angle of the slope is θ?

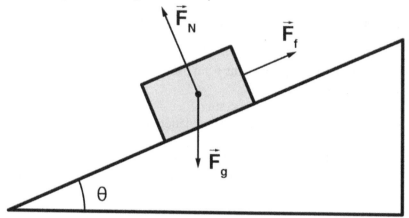

   a. $\vec{F_N} = \cos\theta \times \vec{F_g}$
   b. $\vec{F_N} = \vec{F_g} - \vec{F_f}$
   c. $\vec{F_N} = \tan\theta \times \vec{F_g}$
   d. $\vec{F_N} = \sin\theta \times \vec{F_f}$

2. For the following mechanical system of pulleys and gears, what is the direction and turn rate of the indicated gear? If gear A rotates at 18 rotations per minute, what is the rotation of gear C?

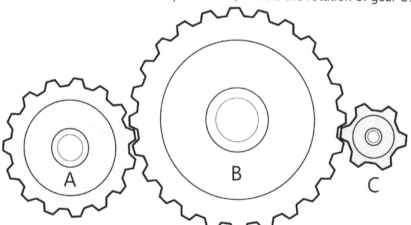

   a.   12 rpm
   b.   54 rpm
   c.   9 rpm
   d.   45 rpm

3. If a driver is traveling west-northwest along a straight road at 70 miles per hour, at what speed are they travelling northwards?
    a. 17.5 mph
    b. 26.8 mph
    c. 53.6 mph
    d. 64.7 mph

4. If a frictionless wooden box at rest is pushed with a constant force of 400 N and its velocity after three seconds is 12 m/s, what is the box's mass in kilograms?
    a. 33.3 kg
    b. 100 kg
    c. 25 kg
    d. 80 kg

5. A ball resting in a person's hand is thrown into the air and falls to the ground. Which of the following statements about the energy of the ball is true, assuming no air resistance?
    a. The potential energy is at its minimum while the ball is resting in the person's hand.
    b. The kinetic energy is at its maximum the instant after the ball is released.
    c. The kinetic energy is greater than the potential energy when the ball has fallen halfway to the ground.
    d. The potential energy is greater resting in the person's hand than on the ground.

6. A box weighing 3 kilograms is stationary on a level, frictionless surface, and another box weighing 1 kilogram is sliding towards it at 5 meters per second. Assuming no kinetic energy is lost, what will be the final speed of the 3-kilogram box?
    a. $5\frac{m}{s}$
    b. $1.67\frac{m}{s}$
    c. $2.5\frac{m}{s}$
    d. $-5\frac{m}{s}$

7. In a partially inelastic collision, which of the following is true?
    a. Momentum and kinetic energy are lost in the collision.
    b. Momentum is conserved but kinetic energy is lost.
    c. Momentum is lost but kinetic energy is conserved.
    d. Momentum and kinetic energy are conserved in the collision.

8. A vehicle begins moving and is continuously accelerating at 3 meters per second squared. After ten seconds, how far has the vehicle traveled?
    a. 30 meters
    b. 45 meters
    c. 75 meters
    d. 150 meters

9. Ships use systems of pulleys called block and tackle systems to aid in moving heavy parts and objects. A block and tackle might wrap a single rope around it multiple times, acting as multiple pulleys. Imagine a rope is attached to the upper block, threaded around the bottom, then to the top and around one more time. Assuming no losses due to friction, what is the ideal mechanical advantage of this setup?

    a. 1
    b. 2
    c. 3
    d. 4

10. A 500-newton box needs to be loaded onto a truck, and a ramp is used to slide it in. If the ramp is four feet high, eight feet long, and the mechanical efficiency of the ramp is 50%, what force is needed to push the box up the ramp?

    a. 100 N
    b. 125 N
    c. 250 N
    d. 87 N

11. Two children, Chris and Elliot, are playing on a seesaw. Chris weighs fifty pounds, and Elliot weighs seventy-five pounds. If the seesaw is twelve feet long and the fulcrum is in the center, where does each child need to sit to balance the seesaw?

    a. Chris should sit 6 feet out, and Elliot 4 feet out.
    b. Chris should sit 5 feet out, and Elliot 3 feet out.
    c. Chris should sit 6 feet out, and Elliot 2 feet out.
    d. Chris should sit 4 feet out, and Elliot 2 feet out.

12. Two weights are suspended on opposite sides of a wheel and axle. A 20-kilogram weight is suspended by a cord wrapped around the axle, and a 10-kilogram weight is suspended from the wheel with its cord wrapped in the opposite direction. Assuming the machine is ideal, if the radius of the axle is half of the wheel's radius, which weight will drop?

    a. Neither; the system is in equilibrium, so the weights will stay still.
    b. The 20-kilogram weight will begin to drop.
    c. The 10-kilogram weight will begin to drop.
    d. There is not enough information to answer this question.

13. A baseball player hits a pitch and sends the ball flying at a 45-degree angle from the ground. When the ball falls back to the same height it began at when it was hit, which of the following will be true, assuming no air resistance?

    a. The ball will have horizontally traveled two times its total vertical distance.
    b. The ball will have traveled equal horizontal and vertical distances.
    c. The ball will have two times more horizontal displacement than vertical.
    d. The ball will have traveled equal horizontal displacement and distance.

14. A frictionless pendulum is swinging in a vacuum. When the pendulum's bob reaches the bottom of its curve, which of the following is true?

    a. The potential energy is at half its maximum.
    b. The kinetic energy is at its minimum.
    c. The potential energy is at its minimum.
    d. The kinetic energy is at its maximum.

15. A 15.0 N sandbag is attached to a rope in the following pulley system. If the pulley system is only two-thirds efficient, what force is needed to suspend the sandbag in the air?

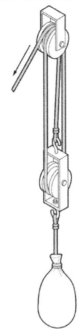

 a. 5.0 N
 b. 7.5 N
 c. 10.0 N
 d. 15.0 N

Assembling Objects

For questions 1–7, which answer choice best shows how the objects on the left will connect if the letters for each object are put together?

1.

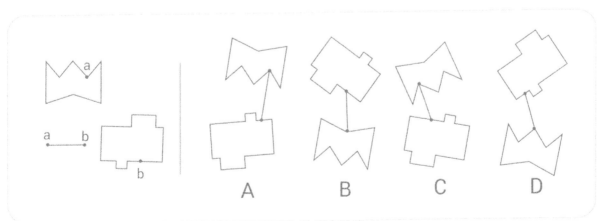

2.

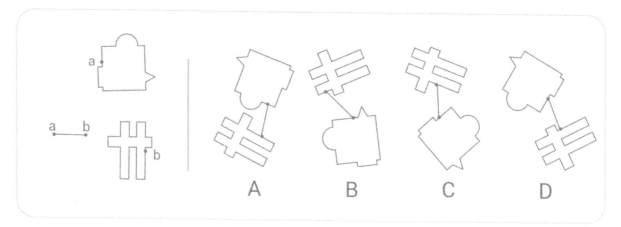

3.

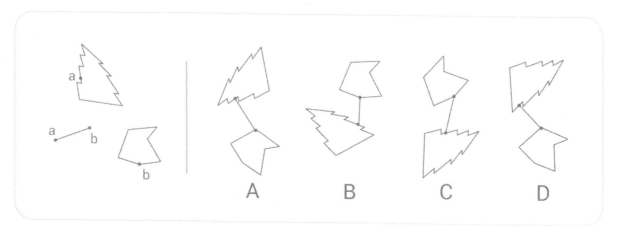

4.

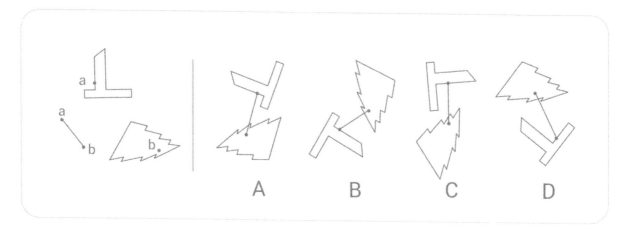

5.

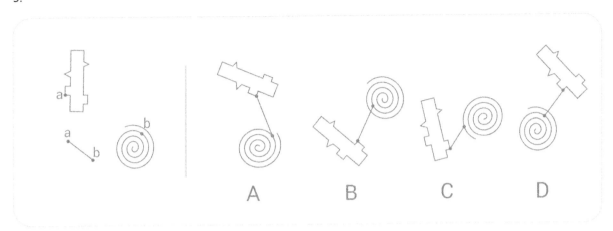

6.

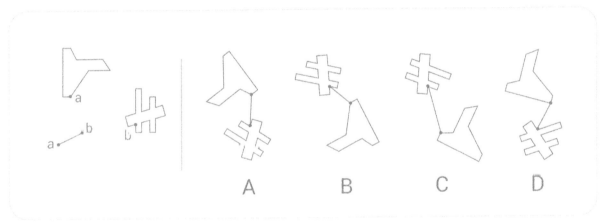

7.

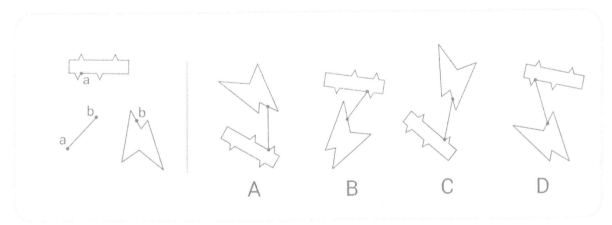

For questions 8–15, which answer choice shows the way the shapes on the left of the image would best fit together?

8.

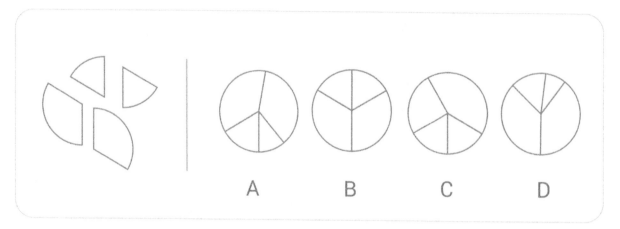

9.

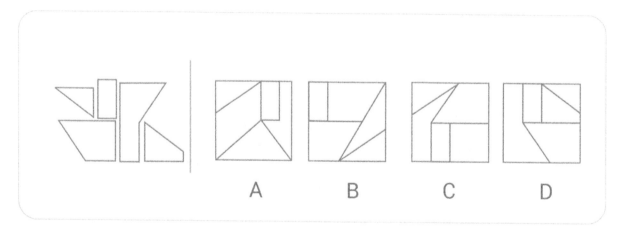

10.

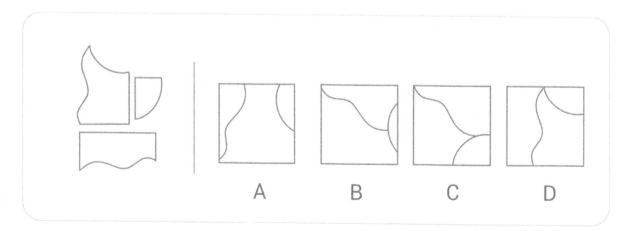

11.

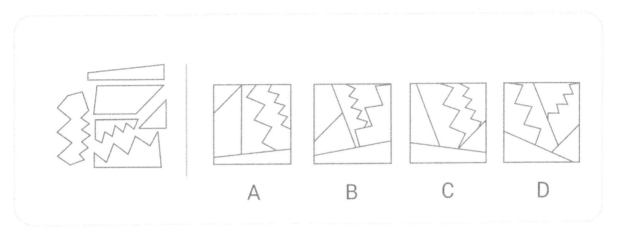

12.

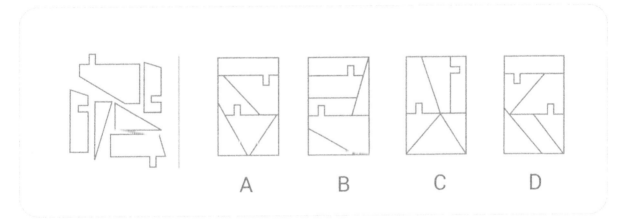

13.

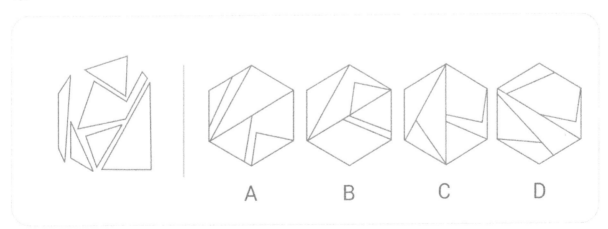

14.

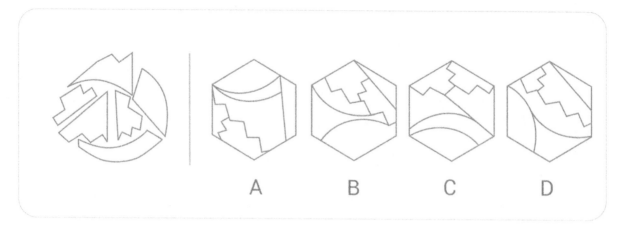

15.

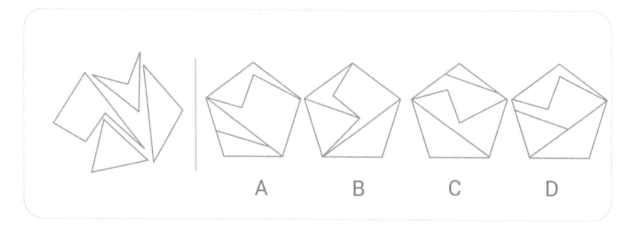

Answer Explanations #1

General Science

1. D: When an object is moving in a circular motion, the force that pushes outward, away from the axis, is called the centrifugal force. It is this outward force that would expel water through the holes in the barrel from the clothes that are pushed up to the sides of the barrel. The centripetal force, Choice *A*, is the force that is pulling the clothes towards the axis. Friction, Choice *B*, does not affect the water being pulled through the holes. Angular momentum, Choice *C*, is a circular force but is not responsible for pulling the water outwards.

2. C: If an object is moving in a negative direction, it has a negative velocity. If the velocity is increasing in that negative direction, it is becoming increasingly more negative and would therefore have a negative acceleration. In Choice *A*, the object would have a positive acceleration. Acceleration is zero in both Choices *B* and *D* because the velocity of the object is not changing.

3. D: After air enters the nose or mouth, it gets passed on to the sinuses. The sinuses regulate temperature and humidity before passing the air on to the lungs. Volume of air can change with varying temperatures and humidity levels, so it is important for the air to be a constant temperature and humidity before being processed by the lungs. The larynx is the voice box of the body, making Choice *A* incorrect. The lungs are responsible for oxygen and carbon dioxide exchange between the air that is breathed in and the blood that is circulating the body, making Choice *B* incorrect. The trachea takes the temperature- and humidity-regulated air from the sinuses to the lungs, making Choice *C* incorrect.

4. A: Chemical reactions are processes that involve the changing of one set of substances to a different set of substances. To accomplish this, the bonds between the atoms in the molecules of the original substances need to be broken. The atoms are rearranged, and new bonds are formed to make the new set of substances. Combination reactions involve two or more reactants becoming one product. Decomposition reactions involve one reactant becoming two or more products. Combustion reactions involve the use of oxygen as a reactant and generally include the burning of a substance. Choices *C* and *D* are not discussed as specific reaction types.

5. C: The main function of the lungs is to provide oxygen to the blood circulating in the body and to remove carbon dioxide from blood that has already circulated the body. The passageways within the lungs are responsible for this gas exchange between the air and the blood. The nose, Choice *A*, is the first place where air is breathed in. The larynx, Choice *B*, is the voice box of the body. The stomach, Choice *D*, is responsible for getting nutrients out of food and drink that are ingested, not from air.

6. A: Monosaccharides are the simplest sugars that make up carbohydrates. They are important for cellular respiration. Fatty acids, Choice *B*, make up lipids. Polysaccharides, Choice *C*, are larger molecules with repeating monosaccharide units. Amino acids, Choice *D*, are the building blocks of proteins.

7. B: Gases are easy to compress. Their molecules spread out to fill the container they are in, but they can be pushed together so that the space between the molecules decreases. Solids have a fixed shape; therefore, Choice *A* is incorrect. Liquids and solids both have an exact volume, so Choice *C* is not correct. Molecules are generally packed together tightly in solids, which makes Choice *D* incorrect.

8. B: White blood cells are part of the immune system and help fight off diseases. Red blood cells contain hemoglobin, which carries oxygen through the blood. Plasma is the liquid matrix of the blood. Platelets help with the clotting of the blood.

9. D: The gastrointestinal system consists of the stomach and intestines, which help process food and liquid so that the body can absorb nutrients for fuel. The respiratory system is involved with the exchange of oxygen and carbon dioxide between the air and blood. The immune system helps the body fight against pathogens and diseases. The genitourinary system refers to the body's urinary and reproductive systems.

10. A: During digestion, complex food molecules are broken down into smaller molecules so that nutrients can be isolated and absorbed by the body. Absorption, Choice *B*, is when vitamins, electrolytes, organic molecules, and water are absorbed by the digestive epithelium. Compaction, Choice *C*, occurs when waste products are dehydrated and compacted. Ingestion, Choice *D*, is when food and liquids first enter the body through the mouth.

11. A: Physical properties of substances are those that can be observed without changing that substance's chemical composition, such as odor, color, density, and hardness. Reactivity, flammability, and toxicity are all chemical properties of substances. They describe the way in which a substance may change into a different substance. They cannot be observed without chemically changing the substance.

12. A: Cellular reproduction is the process of a cell making new cells with similar or the same contents as themselves. As the organism grows, it produces more cells and allows for replacement of cells that have aged or are damaged. It does not mark the end of the organism's life, Choice *B*, or mark the end of growth, as it must grow to reproduce, Choice *C*. Cellular reproduction does not always introduce changes to an organism's DNA; this occurs only if mutations occur. Therefore, Choice *D* is incorrect.

13. C: A reactant is an oxidative agent when it gains electrons, gains an oxygen, or loses a hydrogen. Choices *A* and *B* describe the opposite situations. Carbon is not involved in the transfer of electrons, or when hydrogen acts as a proton, in redox reactions. Therefore, Choice *D* is incorrect.

14. D: The oxidation number of a compound is the charge that the compound would have if it was composed of ions. Neutral compounds do not have a charge, so their oxidation number is 0. The oxidation number of hydrogen is 1, but it is −1 when it is combined with elements that are not as electronegative. Group 2 elements in a compound have an oxidation number of 2. Thus, Choices *A*, *B*, and *C* are incorrect.

15. A: Beyond the visible spectrum of light are shorter wavelengths of light that can be harmful to humans. The Sun has a small percentage of UV light, and when that crosses into the Earth's atmosphere, it can be harmful to human skin and cause burns. The whole visible spectrum, Choice *B*, including blue light, Choice *D*, is not harmful to human skin. Radio waves have long waves, and although they are not visible to the human eye, they are not harmful to the skin; therefore, Choice *C* is incorrect.

Arithmetic Reasoning

1. B: First, the information is translated into the ratio $\frac{15}{80}$. To find the percentage, translate this fraction into a decimal by dividing 15 by 80. The corresponding decimal is 0.1875. Move the decimal point two places to the right to obtain the percentage 18.75%.

2. C: Gina answered 60% of 35 questions correctly; 60% can be expressed as the decimal 0.60. Therefore, she answered $0.60 \times 35 = 21$ questions correctly.

3. B: The unknown quantity is the number of total questions on the test. Let x be equal to this unknown quantity. Since 75% is equal to $\frac{3}{4}$ as a fraction, $\frac{3}{4}x = 12$. To solve for x, multiply both sides by $\frac{4}{3}$. Therefore,

$$x = 12 \times \frac{4}{3} = \frac{48}{3} = 16$$

4. B: If sales tax is 7.25%, the price of the car must be multiplied by 1.0725 to account for the additional sales tax. This is the same as multiplying the initial price by 0.0725 (the tax) and adding that to the initial cost. Therefore:

$$15{,}395 \times 1.0725 = 16{,}511.1375$$

This amount is rounded to the nearest cent, which is $16,511.14.

5. A: Rounding can be used to find the best approximation. All the values can be rounded to the nearest thousand. 15,412 SUVs can be rounded to 15,000. 25,815 station wagons can be rounded to 26,000. 50,412 sedans can be rounded to 50,000. 8,123 trucks can be rounded to 8,000. Finally, 18,312 hybrids can be rounded to 18,000. The sum of the rounded values is 117,000, which is closest to 120,000.

6. D: There are 52 weeks in a year, and if the family spends $105 each week, that amount is close to $100. A good approximation is $100 a week for 50 weeks, which is found through the product:

$$50 \times \$100 = \$5{,}000$$

7. C: This problem involves ratios and proportions. If 12 packets are needed for every 5 people, this statement is equivalent to the ratio $\frac{12}{5}$. The unknown amount x is the number of ketchup packets needed for 60 people. The proportion $\frac{12}{5} = \frac{x}{60}$ must be solved. Cross-multiply to obtain $12 \times 60 = 5x$. Therefore, $720 = 5x$. Divide each side by 5 to obtain $x = 144$.

8. D: There were 48 total bags of apples sold. If 9 bags were Granny Smith and the rest were Red Delicious, then $48 - 9 = 39$ bags were Red Delicious. Therefore, the ratio of Granny Smith to Red Delicious is $9 : 39$.

9. B: The average rate of change is found by calculating the difference in dollars over the elapsed time. Therefore, the rate of change is equal to ($4,900 − $4,000) ÷ 3 months, which is equal to $900 ÷ 3, or $300 per month.

10. A: The formula for the rate of change is the same as slope: change in y over change in x. The y-value in this case is percentage of smokers and the x-value is year. The change in percentage of smokers from 2000 to 2015 was 8.1%. The change in x was $2000 - 2015 = -15$. Therefore, $\frac{8.1\%}{-15} = -0.54\%$. The percentage of smokers decreased 0.54% each year.

11. A: A proportion should be used to solve this problem. The ratio of tagged to total deer in each instance is set equal to one another, and the unknown quantity is a variable x. The proportion is:

$$\frac{300}{x} = \frac{5}{400}$$

Cross-multiplying gives $120,000 = 5x$, and dividing through by 5 results in 24,000.

12. B: The number of representatives varies directly with the population, so the equation necessary is $N = k \times P$, where N is number of representatives, k is the variation constant, and P is total population in millions. Plugging in the information for New York allows k to be solved for. This process gives $27 = k \times 20$, so $k \approx 1.35$. Therefore, the formula for number of representatives given total population in millions is:

$$N = 1.35 \times P$$

Plugging in $P = 11.6$ for Ohio results in $N \approx 15.66$, which rounds up to 16 total representatives.

13. B: Of the 60 stores, 45 are clothing stores. Therefore, $0.45 \times 60 = 27$, so 27 stores are clothing stores. The rest are not clothing stores: $60 - 27 = 33$. Therefore, 33 stores are not clothing stores.

14. B: To find the hourly rate of the van, divide the total cost by the number of hours Steve rented it. The total number of hours was 9 (9 a.m. to 6 p.m.). Therefore, the hourly rate of the van was:

$$\frac{\$140.85}{9} = \$15.65$$

15. C: From Katy's hourly wage, she earned $\$15(20) = \$300$. From commission, she earned 6% of $2,400, or $0.06(2,400) = \$144$. Therefore, she earned $\$300 + \$144 = \$444$ total that week.

Word Knowledge

1. D: Someone who is *wary* is cautious or apprehensive. This word is often used in the context of being watchful or on guard about a potential danger. For example, darkening clouds and white caps on the waves may make a seaman wary about setting sail.

2. D: *Proximity* is defined as closeness, or the state or quality of being near in place, time, or relation.

3. C: As an adjective, *parsimonious* means frugal to the point of being stingy, or very unwilling to spend money, which is similar to being miserly.

4. B: The noun *propriety* describes something or someone as being suitable or appropriate for the circumstances or purpose. It can mean conformity to accepted standards, particularly as they relate to good behavior or manners. In this way, *propriety* can be considered to mean the state or quality of being proper. For example, beyond simply upholding the law, Americans typically expect their president to act with propriety.

5. C: A *boon* is a benefit or blessing, often considered to be timely. It is something to be thankful for. For example, a new tax benefit enacted for first-time homeowners would be a boon to a family who just closed on a house. In an alternative usage, it can be a favor or a benefit given upon request.

6. D: *Strife* is a noun that is defined as bitter or vigorous discord, conflict, or dissension. It can mean a fight or struggle, or other act of contention. For example, antagonistic political interest groups vying for local support may be at strife.

7. B: *Qualm* is a noun that means a feeling of apprehension or uneasiness, often brought on suddenly. A girl who is just learning to ride a bike may have qualms about getting back on the saddle after taking a bad fall. It may also refer to an uneasy feeling related to one's conscience as it pertains to their actions. For example, a man with poor morals may have no qualms about lying on his tax return.

8. D: *Valor* is bravery or courage when facing a formidable danger. It often relates to strength of mind or spirit during battle or acting heroically in such situations.

9. C: *Zeal* is eagerness, fervor, or ardent desire in the pursuit of something. For example, a competitive collegiate baseball player's zeal to succeed in his sport may compromise his academic performance.

10: A: To *extol* is to highly praise, laud, or glorify. People often extol the achievements of their heroes or mentors.

11. B: To *reproach* means to blame, find fault, or severely criticize, sometimes for the purpose of correcting behavior. Chide similarly means to voice disapproval or criticize in a constructive manner. Reproach can also be defined as expressing significant disapproval. It is often used in the phrase "beyond reproach" as in, "her violin performance was beyond reproach." In this context, it means her playing was so good that it evaded any possibility of criticism.

12. D: *Milieu* refers to the social surroundings or environment of a person.

13. D: *Guile* can be defined as the quality of being cunning or crafty and skilled in deception. Someone may use guile to trick or deceive somebody into giving them money.

14. A: As a verb, *assent* means to express agreement, give consent, or acquiesce. A job candidate might assent to an interviewer's request to perform a background check. As a noun, it means an agreement, acceptance, or acquiescence.

15. B: *Dearth* means a lack of something, or a scarcity or shortage. For example, a local library might have a dearth of information pertaining to an esoteric topic.

Paragraph Comprehension

1. A: Choice *B* is incorrect because although it is obvious the author favors rehabilitation, the author never asks for donations from the audience. Choices *C* and *D* are also incorrect. We can infer from the passage that American prisons are probably harsher than Norwegian prisons. However, the best answer that captures the author's purpose is Choice *A*, because the author compares Norwegian and American prison recidivism rates.

2. C: The narrator, after feeling excitement for the morning, recalls an interaction with Peter Walsh when she was 18. She states that on that day she had felt "that something awful was about to happen," which is another way to say that she had a sense of foreboding. The narrator mentions larks and weather in the passage, but there is no proof of anger or confusion at either of them.

3. B: To convince the audience that judges holding their positions based on good behavior is a practical way to avoid corruption. Choice *A* is incorrect because although he mentions the condition of good behavior as a barrier to despotism, he does not discuss it as a practice in the states. Choice *C* is incorrect because the author does not argue that the audience should vote based on judges' behavior, but rather that good behavior should be the condition for holding their office. Choice *D* is not represented in the passage, so it is incorrect.

4. B: Whatever happened in his life before he had a certain internal change is irrelevant. Choices *A, C,* and *D* use some of the same language as the original passage, like "revolution," "speak," and "details," but they are incorrect. The narrator is not concerned with the character's life before his epiphany and has no intention of talking about it.

5. C: A certain characteristic or quality imposed upon something else. The sentence states that "the impress of his dear hand [has] been stamped everywhere." Choice *A* is one definition of *impress*, but this definition is used more as a verb than a noun: "She impressed us as a songwriter." Choice *B* is incorrect because it is also used as a verb: "He impressed the need for something to be done." Choice *D* is incorrect because it is part of a physical act: "the businessman impressed his mark upon the envelope." The phrase in the passage is meant as figurative, since the workmen did most of the physical labor, not the prince.

6. B: Mr. Ford assumed Booth's movement throughout the theater was due to being familiar with the theater. Choice *A* is incorrect; although Booth does eventually make his way to Lincoln's box, Mr. Ford does not make this distinction in this part of the passage. Choice *C* is incorrect; although the passage mentions "companions," it mentions Lincoln's companions rather than Booth's companions. Finally, Choice *D* is incorrect; the passage mentions "dress circle," which means the first level of the theater, but this is different from a "dressing room."

7. C: The tone of this passage is neutral since it is written in an academic/informative voice. It is important to look at the author's word choice to determine what the tone of a passage is. We have no indication that the author is excited, angry, or sorrowful at the effects of bacteria on milk, so Choices *A, B,* and *D* are incorrect.

8. D: A city whose population is made up of people who seek quick fortunes rather than building a solid business foundation. Choice *A* is a characteristic of Portland, but not of a boom city. Choice *B* is close—a boom city is one that becomes quickly populated, but it is not necessarily always populated by residents from the east coast. Choice *C* is incorrect because a boom city is not one that catches fire frequently, but one made up of people who are looking to make quick fortunes from the resources provided on the land.

9. B: Look for the key words that give away the type of passage this is, such as *deities, Greek hero, centaur,* and the names of demigods like Artemis. A eulogy is typically a speech given at a funeral, making Choice *A* incorrect. Choices *C* and *D* are incorrect, as *virgin huntresses* and *centaurs* are typically not found in historical or professional documents.

10. B: Although we don't have much context to go off of, this person is probably not a "lifetime partner" or "farm crop" of civilized life. These two do not make sense, so Choices *C* and *D* are incorrect. Choice *A* is also a bit strange. To be an "illegal immigrant" of civilized life is not a used phrase, making Choice *A* incorrect.

Mathematics Knowledge

1. C: By switching from a radical expression to rational exponents:

$$\sqrt[4]{x^6} = x^{\frac{6}{4}} = x^{\frac{3}{2}}$$

Also, properties of exponents can be used to simplify $\frac{x}{x^3}$ into:

$$x^{1-3} = x^{-2} = \frac{1}{x^2}$$

The other terms can be left alone, resulting in an equivalent expression:

$$x^{\frac{3}{2}} - \frac{1}{x^2} + x - 2$$

2. B: Using Descartes' Rule of Signs, count the number of sign changes in coefficients in the polynomial. This results in the number of possible positive zeros. The coefficients are 1, −3, 2, 1, and −3, so the sign changes from 1 to −3, −3 to 2, and 1 to −3, a total of 3 times. Therefore, there are at most 3 positive zeros.

3. B: The goal is to first isolate the variable. The fractions can easily be cleared by multiplying the entire inequality by 5, resulting in $35 - 4x < 3$. Then, subtract 35 from both sides and divide by −4. This results in $x > 8$. Notice the inequality symbol has been flipped because both sides were divided by a negative number. The solution set, all real numbers greater than 8, is written in interval notation as $(8, \infty)$. A parenthesis shows that 8 is not included in the solution set.

4. D: Let x be the unknown number. Difference indicates subtraction, and sum represents addition. To triple the difference, it is multiplied by 3. The problem can be expressed as the following equation:

$$3(5 - x) = x + 5$$

Distributing the 3 results in:

$$15 - 3x = x + 5$$

Subtract 5 from both sides, add $3x$ to both sides, and then divide both sides by 4. This results in:

$$x = \frac{10}{4} = \frac{5}{2} = 2.5$$

5. B: The expression under the radical can be factored into perfect squares as such:

$$\sqrt{100 \times 2 \times x^6 \times y^6 \times y \times z^2}$$

Using rational exponents, this is the same as

$$(100 \times 2 \times x^6 \times y^6 \times y \times z^2)^{\frac{1}{2}}$$

$$100^{\frac{1}{2}} \times 2^{\frac{1}{2}} \times x^{\frac{6}{2}} \times y^{\frac{6}{2}} \times y^{\frac{1}{2}} \times z^{\frac{2}{2}}$$

$$\sqrt{100} \times \sqrt{2} \times x^3 \times y^3 \times \sqrt{y} \times z$$

$$10x^3y^3z\sqrt{2y}$$

6. D: The exponential rules $(ab)^m = a^m b^m$ and $(a^m)^n = a^{mn}$ can be used to rewrite the expression as:

$$4^4 y^{12} \times 3^2 y^{14}$$

The coefficients are multiplied together and the exponential rule $a^m a^n = a^{m+n}$ is then used to obtain the simplified form $2{,}304y^{26}$.

7. A: The surface area for a cylinder is the sum of the two circle bases and the rectangle formed on the side. This is easily seen in the net of a cylinder.

The Net of a Cylinder

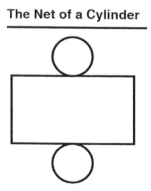

The area of a circle is found by multiplying π times the radius squared. The rectangle's area is found by multiplying the circumference of the circle by the height. The equation $SA = 2\pi \times 5 \times 10 + 2\pi \times 5^2$ shows the area of the rectangle as $2\pi \times 5 \times 10$, which yields 314. The area of the bases is found by $\pi \times 5^2$, which yields 78.5, and then is multiplied by 2 for the two bases. Adding these together gives 471.

8. A: The area of the shaded region is calculated in a few steps. First, the area of the rectangle is found using the formula:

$$A = length \times width = 6\,\text{m} \times 2\,\text{m} = 12\,\text{m}^2$$

Second, the area of the triangle is found using the formula:

$$A = \frac{1}{2} \times base \times height = \frac{1}{2} \times 3\,\text{m} \times 2\,\text{m} = 3\,\text{m}^2$$

The last step is to take the rectangle area and subtract the triangle area. The area of the shaded region is:

$$A = 12\,\text{m}^2 - 3\,\text{m}^2 = 9\,\text{m}^2$$

9. B: The fraction $\frac{12}{60}$ can be reduced to $\frac{1}{5}$, in lowest terms. First, it must be converted to a decimal. Dividing 1 by 5 results in 0.2. Then, to convert to a percentage, move the decimal point two places to the right and add the percentage symbol. The result is 20%.

10. B: Multiply both the numerator and denominator by 3. This results in the fraction $\frac{42}{99}$. Since we are rounding, we can make this the fraction $\frac{42}{100}$, which can be converted to a decimal of 0.42.

11. B: Common denominators must be used. The LCD is 15, and $\frac{2}{5} = \frac{6}{15}$. Therefore, $\frac{14}{15} + \frac{6}{15} = \frac{20}{15}$, and in lowest terms, the answer is $\frac{4}{3}$. A common factor of 5 was divided out of both the numerator and denominator.

12. A: A product is found by multiplication. Multiplying two fractions together is easier when common factors are canceled first to avoid working with larger numbers.

$$\frac{5}{14} \times \frac{7}{20} = \frac{5}{2 \times 7} \times \frac{7}{5 \times 4} = \frac{1}{2} \times \frac{1}{4} = \frac{1}{8}$$

13. D: Division is completed by multiplying by the reciprocal:

$$24 \div \frac{8}{5} = \frac{24}{1} \times \frac{5}{8} = \frac{3 \times 8}{1} \times \frac{5}{8} = \frac{15}{1} = 15$$

14. C: Common denominators must be used. The LCD is 168, which can be found by finding the LCM of 14 and 24, so each fraction must be converted to have 168 as the denominator.

$$\frac{5}{24} - \frac{5}{14} = \frac{5}{24} \times \frac{7}{7} - \frac{5}{14} \times \frac{12}{12} = \frac{35}{168} - \frac{60}{168} = -\frac{25}{168}$$

15. C: The correct mathematical statement is the one in which the number to the left on the number line is less than the number to the right on the number line. It is written in Choice C that $\frac{1}{3} > -\frac{4}{3}$, which is the same as $-\frac{4}{3} < \frac{1}{3}$, a correct statement.

Electronics Information

1. B: The total resistance in a series circuit is equal to the sum of all resistances because the current must pass through each resistive element, so each resistance applies to the circuit.

2. C: The only true relationship shown is $PR = V^2$. Ohm's law relates current, voltage, and resistance as $V = IR$, and power is the rate of movement of a charge across an electric potential difference, or $P = \frac{q}{t} \times \frac{E}{q} = IV$.

3. C: The point with the least current is through R_3. Even without knowing the voltage or total current, it's possible to find relative values in the circuit. Recall that current is divided through parallel branches, inversely proportional to their resistance, but is the same at all points in series. Consequently, the total current, I, across R_1, R_2, and R_3 is equal to the total current across R_4 and R_5. Additionally, for branches of equal resistance, current is divided equally. So, the current through R_4 and R_5 may each be represented as $\frac{1}{2}I$. If the resistances of R_1, R_2, and R_3 were equal, the current would be $\frac{1}{3}I$ through each—less than through R_4 and R_5. However, because the voltage across each parallel resistor is equal and R_3 has a greater resistance than R_1 or R_2, less current flows through it. Thus, R_3 is the point with the least current.

4. A: If power decreases and resistance stays the same, the product of voltage and current must also decrease, because power is equal to the product of current and voltage, $P = IV$. Choices B and C are incorrect because if one increases and the other remains constant, power would increase. Choice D is incorrect because if both current and voltage increase, power would also increase.

5. C: The voltage between points A and B is 3 volts. The voltage between any two points in series can be found by multiplying current by the resistance between the two points. The current in series is found by taking the total voltage (9 volts) and dividing it by the sum of all resistances (13.5 kiloohms), which yields $\frac{2}{3}$ mA, or approximately 0.667 mA:

$$\frac{9 \text{ V}}{13,500 \text{ }\Omega} = \frac{2}{3,000} \text{ A} = 0.667 \text{ mA}$$

The current is then multiplied by 4.5 kiloohms to find the voltage between points A and B:

$$0.667 \text{ mA} \times 4.5 \text{ k}\Omega \approx 3 \text{ V}$$

The electric potential difference between A and B is approximately 3 volts.

6. B: The power source provides 120 volts. The current of 0.5 amps is given, so taking the sum of resistances, $180 \text{ }\Omega + 60 \text{ }\Omega = 240 \text{ }\Omega$, and multiplying it by current yields:

$$0.5 \text{ A} \times 240 \text{ }\Omega = 120 \text{ V}$$

The supply's voltage is 120 volts.

7. D: Flow of charges in a circuit is through the movement of electrons, relatively small particles which carry a negative charge. Choices *A* and *C* are incorrect because the protons in conductors are not a circuit's charge carriers. Choice *B* is incorrect because the direction of electron flow is from a negatively charged location toward the positive, because like charges repel each other.

8. A: This question can be solved using the relationship of power to energy and Ohm's law. Power is equal to energy over time, $P = \frac{E}{t}$; consequently, energy is also equal to the product of power and time, $E = P \times t$. Power is also equal to voltage times current, and current is equal to charge over time:

$$P = VI = V \times \frac{q}{t}$$

Substituting this in for power in the first equality yields:

$$E = P \times t = V \times \frac{q}{t} \times t = V \times q$$

Therefore, the product of voltage and charge is equivalent to energy.

9. B: The voltage drop across a point in parallel can be found by calculating the equivalent resistance of the parallel branches and multiplying it by the total current, because of the electrically common points before and after parallel branches. The equivalent resistance of R_1 and R_2 is:

$$R_{eq} = \frac{1}{\frac{1}{1 \text{ k}\Omega} + \frac{1}{2.2 \text{ k}\Omega}} = \frac{1}{1.45 \text{ k}\Omega} = 0.69 \text{ k}\Omega$$

Then the current may be calculated using the total voltage and sum of series resistances:

$$I_T = \frac{V_T}{R_T} = \frac{12 \text{ V}}{10 \text{ k}\Omega + 0.69 \text{ k}\Omega} = \frac{12 \text{ V}}{10.69 \text{ k}\Omega} = 1.12 \text{ mA}$$

Because the voltage is equivalent between points sharing a common point, the voltage drop across R_2 is equal to the voltage drop across the equivalent resistance, R_{eq}:

$$V_{eq} = I_T \times R_{eq} = 1.12\text{ mA} \times 0.69\text{ k}\Omega = 0.77\text{ V}$$

Thus, the voltage drop across R_2 is 0.77 V.

10. A: Joules are a measure of energy. Choice *B*, watts, is incorrect because it represents power, or energy over time. Choice *C*, coulombs, is incorrect because it is a quantity of electric charge. Choice *D*, amperes, is incorrect because it is the unit for current, which is the rate of charges moved through a point, or charge over time.

11. C: In an open circuit, there is in fact significantly *more* resistance, because air and most other materials are not as conductive as the conductor material typically used in circuits, such as copper or aluminum. Materials that are less conductive are *more resistive*, because resistivity is the inverse of conductivity.

12. D: The current through I in the circuit is 4 amps. Kirchoff's current law states that the sum of all currents toward a junction is equal to the sum of all currents exiting the same point. In the diagram provided, the current moving toward the junction immediately left of I adds up to 10 amps. The current moving away from the same point is 6 amps and I. If these are summed together, the current through I can be solved:

$$8\text{ A} + 2\text{ A} = I + 6\text{ A}$$

$$I = 10\text{ A} - 6\text{ A} = 4\text{ A}$$

This can alternatively be solved using the junction on the right of I. The current moving toward that junction is 3 A and I, and the current away is 7 A:

$$I + 3\text{ A} = 7\text{ A}$$

$$I = 7\text{ A} - 3\text{ A} = 4\text{ A}$$

Thus, the current at I is 4 amps.

13. B: The question asks what the unknown resistance must be if the voltage is zero, or in other words, if the electric potential at both points is equal. The resistance of R_x must balance out the ratio of the known resistances. The resistances may be set up as a proportion:

$$\frac{R_x}{R_3} = \frac{R_2}{R_1}$$

$$\frac{R_x}{1.6\ \Omega} = \frac{1.3\ \Omega}{2.2\ \Omega} \approx 0.9\ \Omega$$

Thus, the unknown resistance R_x is 0.9 ohms.

14. A: Resistance is inversely proportional to conductance. Mathematically, their relationship is represented as $R = \frac{1}{C}$ or $C = \frac{1}{R}$. Impedance is related to resistance because it includes resistance and reactance. Reactance is a property of AC circuits which resists flow of current but does not lose energy as heat like resistance.

15. A: The area of the capacitor plates is 1.30 millimeters. Recall the equation for capacitance of parallel plate capacitor:

$$C = \kappa \frac{\varepsilon_0 A}{d}$$

Capacitance, in this case, is already known. The unknown is the area of the plates. The equation can be rearranged algebraically to give:

$$A = \frac{C \times d}{\kappa \times \varepsilon_0}$$

Then, substitute in the known values for capacitance, distance, relative permittivity, and the vacuum permittivity constant ($8.85 \times 10^{-1} \frac{F}{m}$):

$$A = \frac{5.3 \text{ pF} \times 5 \text{ mm}}{2.3 \times 8.85 \times 10^{-12} \frac{F}{m}} = \frac{5.3 \times 10^{-12} \text{ F} \times 0.005 \text{ m}}{2.3 \times 8.85 \times 10^{-12} \frac{F}{m}} = 0.00130 \text{ m}^2 = 1.30 \text{ mm}^2$$

Thus, the area of the plates is 1.30 mm².

Auto Information

1. C: The spark plug activates as the piston finishes extending at the end of the compression stroke. This is the point when fuel and air are close together and near the plug to maximize the likelihood of ignition. When the intake valve opens, that's the start of the intake stroke, the influx of fuel-air mixture to the cylinder, so Choice *A* is incorrect. Choice *B* is incorrect because that is the beginning of the compression stroke.

2. B: A dead blow hammer is the ideal choice for dislodging hardware because it has shifting internal weights to reduce bouncing after striking a surface, thus minimizing the risk of damaging other parts. Choice *A*, needle nose pliers, are designed for holding, bending, and otherwise manipulating smaller objects. Choice *C*, a vise grip, is designed to grab hardware without slipping or to hold multiple pieces together. Choice *D*, a ball-peen hammer, is a hammer with a rounded head to strike and reshape metal surfaces.

3. D: The ignition coil is an inductor that is used to increase the voltage from the battery high enough to create a spark. The battery voltage, depending on its chemical makeup and wear, remains around the same value. The distributor is designed to redirect the voltage from the ignition coil to the correct spark plug, while the ignition control module is used to control the timing of the ignition system.

4. C: The drive shaft, although it connects indirectly to the wheels, is not responsible for braking. Choices *A* and *B* are incorrect because they are both part of the brake system; brake calipers are actuated to force the brake pads against a brake rotor, while brake drums provide a surface for brake shoes to clamp against. Choice *D* is also incorrect because the master cylinder takes force from the brake pedal and conveys it to the brake lines.

5. B: Nitrogen dioxide is a relatively benign emission. The catalytic converter in the exhaust system is designed to convert damaging emissions (carbon monoxide, nitric oxide, and evaporated gas) into less harmful combustion products such as oxygen, nitrogen dioxide, carbon dioxide, and water.

6. D: Ancillary circuits are parts of the vehicle's electrical system that come directly from the battery and do not require the ignition switch being on. The car's radio, lights, and starter engine do not require the vehicle to be started.

7. D: The differential is a structural component on the vehicle, sturdy enough and designed to be able to hold the weight of the vehicle (as it is connected to the axles). Although the tires normally hold the weight of the vehicle, they are not reliable as lift points for jacks and lifts. The engine and driveshaft are similarly not used as lift points.

8. A: A P0300 diagnostic code is a standardized error code for the engine's ignition system misfiring. Without memorizing every code, the problem can be solved by looking at the first two characters of the code. The letter "P" is used for powertrain issues, which includes the engine, and the second character indicates whether it is a standard or manufacturer-specific error. Choice *A* is the only code that has both.

9. B: The transmission fluid, although an oily fluid used to lubricate the transmission, is also a hydraulic fluid specifically part of the transmission system, used to transmit force. The oil sump holds all the extra oil for the engine and the feed lines transmit it. The oil pump (sump pump), located near the bottom of the oil sump, pushes engine oil up through the feed lines.

10. C: The crankshaft converts the linear motion of the pistons into rotational motion, providing the driving force of the vehicle, so the answer is Choice *C*. The other choices all convert different types of motion. The differential is the part of the transmission that enables the left and right wheels to rotate at different rates during a turn, while the camshaft does the opposite of the crankshaft—it converts rotational motion into reciprocal linear motion to open and close the engine valves. The gearbox increases or decreases torque by changing the speed of its output.

Shop Information

1. A: A hacksaw is most commonly used to cut metal pipes. Choices *B*, *C*, and *D* are incorrect because they are all used in different ways to cut wood.

2. B: A face shield should be used to protect the face from flying debris. Choices *A* and *C* are incorrect because gloves and aprons are not used to protect the face. Choice *D* is incorrect because face masks are used to protect the wearer from inhaling harmful particles rather than from facial injury.

3. C: Using a hammer and chisel together allows a worker, such as a sculptor, to cut and shape various materials to make designs and shapes or alter the material's surface. Choice *A* is incorrect because nails are used to fasten materials together. Choice *B* is incorrect because wedges are designed to separate materials, or to secure them using outward pressure. Choice *D* is incorrect because a handsaw is used to cut wood by itself, not in conjunction with a hammer.

4. B: A dead blow hammer, filled with sand or some other loose substrate, is designed for minimal recoil and surface damage. Choices *A*, *C*, and *D* are incorrect, as these are fictitious tool names that do not exist.

5. A: An exhaust hood is used to capture and remove harmful inhalants from the workshop. Choice *B* is incorrect because covering the exhaust fan would make the ventilation system ineffective. Choice *C* is incorrect because an exhaust hood does not minimize noise. Choice *D* is incorrect because collection of contaminant particles would be the purpose of a filter.

6. D: The image shows a vice. It is sometimes known as a bench vice because it is normally attached to a workbench. Choices *A*, *B*, and *C* are incorrect because pliers, clamps, and wrenches are handheld tools that can be moved for flexibility of use.

7. C: A planer is used to cut away thin strips from the surface of a piece of wood. Choice *A* is incorrect because a drill is used to make holes in pieces of wood and other materials. Choice *B* is incorrect because a bandsaw would not normally be used to cut away thin strips of wood; it is better suited for making large cuts. Choice *D* is incorrect because a level is a tool designed to measure how angled a surface is.

8. B: A rip saw is used to make cuts along the grain of wood. Choice *A* is incorrect because fret saws are used to make precise, intricate cuts into wood. Choice *C* is incorrect because a crosscut saw is designed to cut across (or perpendicular to) the grain. Choice *D* is incorrect because a hacksaw is normally used to cut pieces of metal, such as pipes.

9. A: An open-end wrench, although it normally has two different-sized ends, is not adjustable to different-sized tasks. Choices *B*, *C*, and *D* are incorrect because crescent wrenches, monkey wrenches, and pliers are all adjustable to different sizes to meet the needs of a variety of situations.

10. C: A wood router can be used to engrave intricate designs into wood. Choice *A* is incorrect because cutting a wooden board is the purpose of a saw. Choice *B* is incorrect because making holes in wooden boards is the purpose of a drill. Choice *D* is incorrect because making smooth edges on wood is the purpose of sandpaper, files, or sanding tools.

Mechanical Comprehension

1. A: The equation for the normal force in this system is $\overrightarrow{F_N} = \cos\theta \times \overrightarrow{F_g}$. The angle between between the slope and the ground is the same as the angle between the normal force (perpendicular to the slope) and gravity. Choice *B* looks close, because the sum of all forces in a static system must be zero. However, subtracting friction from the weight results in a non-zero net force. The other two choices do not use the angle correctly to find the normal force.

2. D: The rotation of gear C is 30 rpm. To find the angular velocity in a gear ratio problem, a proportion can be set up. Omega (ω) represents angular velocity:

$$\omega_A \times \text{number of teeth of gear A} = \omega_B \times \text{number of teeth of gear B}$$

$$\frac{\omega_A \times \text{number of teeth of gear A}}{\text{number of teeth of gear B}} = \omega_B$$

This can be repeated for gears B and C:

$$\frac{\omega_B \times \text{number of teeth of gear B}}{\text{number of teeth of gear C}} = \omega_C$$

The first equation can be substituted into the second in place of ω_B to find the ratio of gear A to gear C:

$$\omega_A \times \frac{N_A}{N_B} \times \frac{N_B}{N_C} = \omega_C$$

The equation cancels out N_B, leaving the ratio of gear A to C. Plug in the values of each variable to solve for ω_C:

$$\omega_A \times \frac{N_A}{N_C} = \omega_C$$

$$18 \text{ rpm} \times \frac{15}{6} = \frac{18 \times 15}{6} = \frac{3 \times 15}{1} = 45 \text{ rpm}$$

3. B: To find the northward speed of the driver, take the sine of the angle traveled and multiply it by the magnitude of the driver's velocity. The driver is traveling west by northwest, meaning they're traveling at an angle halfway between west and northwest, meaning 25% of 90 degrees or 22.5 degrees north of west. The sine of this angle gives a factor corresponding to the northwards direction traveled, approximately 0.38. When this is multiplied by the total velocity, it yields:

$$\sin(90 \times 0.25) \times 70\,\frac{\text{mi}}{\text{h}} = \sin(22.5) \times 70\,\frac{\text{mi}}{\text{h}} \approx 26.8\,\frac{\text{mi}}{\text{h}}$$

4. B: To find the mass of the box, use the force equation, $F = ma$. The force is known to be 400 newtons, so acceleration has to be found. The box starts at 0 m/s and accelerates to 12 m/s over 3 seconds, meaning its acceleration is 4 m/s². Plug in the known values, and solve for m:

$$F = ma$$

$$400 \text{ N} = m\left(\frac{12\,\frac{\text{m}}{\text{s}}}{3\,\text{s}}\right)$$

$$400 \text{ N} = m\left(4\,\frac{\text{m}}{\text{s}^2}\right)$$

Recall that 1 newton is 1 kilogram-meter per second squared.

$$m = \frac{400\,\frac{\text{kg} \times \text{m}}{\text{s}^2}}{\left(4\,\frac{\text{m}}{\text{s}^2}\right)} = 100 \text{ kg}$$

Thus, the mass of the box is 100 kilograms.

5. D: The potential energy of the ball is higher while in the person's hand than on the ground. Recall the equations for potential energy and kinetic energy, $PE = mgh$ and $KE = \frac{1}{2}mv^2$. For an object of a given mass, potential energy is proportional to the height of the object, while kinetic energy is proportional to the square of its velocity. An object at rest has zero velocity, and thus no kinetic energy, so all its energy must be stored as potential energy. The minimum potential energy of the ball will be when it rests on the ground, so Choice *A* is incorrect. The kinetic energy will be at its maximum when the ball reaches its maximum velocity. Since air resistance is being ignored, and the ball is falling to the ground rather than back to the person's hand, it will be at its maximum velocity the instant before it hits the ground, so Choice *B* is also incorrect. When the ball is at half of its maximum height, its potential energy is at half. Since the total energy remains unchanged, the kinetic energy must also be half, so they are equal, and Choice *C* is incorrect.

6. C: Since the collision is completely elastic, both momentum and kinetic energy will be conserved, and thus, the following pair of equations can be used:

$$v_{1f} = \left(\frac{m_1 - m_2}{m_1 + m_2}\right) v_{1i} + \left(\frac{2 \times m_2}{m_1 + m_2}\right) v_{2i}$$

If we treat the stationary 3-kilogram box as m_1 and the 1-kilogram box as m_2, and their initial velocities as v_{1i} and v_{2i} respectively, then the equation can be simplified and solved for v_{1f}. Because the 3-kilogram box has no initial velocity, the first portion of the equation is equal to zero and can be removed:

$$v_{1f} = \left(\frac{2 \times m_2}{m_1 + m_2}\right) v_{2i}$$

Then, plugging in the known values, we get:

$$v_{1f} = \left(\frac{2 \times 1\,\text{kg}}{3\,\text{kg} + 1\,\text{kg}}\right)\left(5\frac{\text{m}}{\text{s}}\right)$$

$$v_{1f} = \left(\frac{2\,\text{kg}}{4\,\text{kg}}\right)\left(5\frac{\text{m}}{\text{s}}\right)$$

$$v_{1f} = \frac{1}{2} \times 5\frac{\text{m}}{\text{s}}$$

$$v_{1f} = 2.5\frac{\text{m}}{\text{s}}$$

Since the sign is the same as the initial velocity of the 1-kilogram box, we know the 3-kilogram box will continue to move in the same direction as the other box, at 2.5 meters per second.

7. B: In any inelastic collision, momentum is conserved, but some amount of kinetic energy is lost as heat and deformation. Choices *A* and *C* are incorrect because collisions obey conservation of momentum. Choice *D* is incorrect because kinetic energy is only conserved in perfectly elastic collisions.

8. D: The vehicle's displacement will be 150 meters after ten seconds. Since the vehicle is accelerating continuously in a line from a standstill, the displacement can be found by the following kinematic equation, where d is displacement, d_0 is initial displacement (in this case, zero), v_0 is initial velocity, a is acceleration, and t is time:

$$d = d_0 + v_0 t + \frac{1}{2} a t^2$$

$$d = 0 + 0\frac{\text{m}}{\text{s}} \times 10\,\text{s} + \frac{1}{2} \times 3\frac{\text{m}}{\text{s}^2} \times (10\,\text{s})^2$$

$$d = 1.5\frac{\text{m}}{\text{s}^2} \times 100\,\text{s}^2 = 150\,\text{m}$$

Thus, the vehicle has moved 150 meters.

9. D: The mechanical advantage of the system is 4. Imagining the system as multiple pulleys can help aid in finding the mechanical advantage:

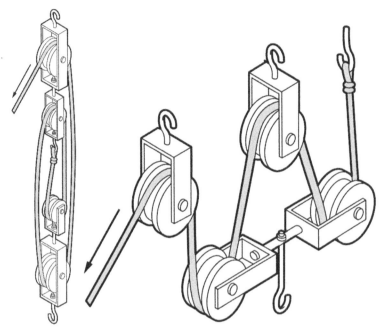

What becomes apparent is that the load is held up by four segments of cable, so the load is split across the four segments and its mechanical advantage is 4. Alternatively, the number of times the cable crosses between the pulley(s) holding the load and the pulleys fixed in place can be counted. It crosses the center four times, which is equal to the ideal mechanical advantage of the system.

10. B: The force needed to push the box is at least 125 newtons. Because an inclined plane effectively reduces the required lifting force by a ratio, the required force can be calculated by multiplying the weight of the box in newtons by the mechanical advantage of the inclined plane and its mechanical efficiency (the ratio of loss due to friction):

$$F = 500 \text{ N} \times \frac{4 \text{ ft}}{8 \text{ ft}} \times 0.5 = 500 \text{ N} \times \frac{1}{2} \times 0.5 = 125 \text{ N}$$

11. A: Chris should sit 6 feet out, and Elliot 4 feet out. The mechanical advantage of a lever is a proportion of the force times distance from the fulcrum for the load and effort. In this case, the children on the seesaw take turns applying force to push off the ground. To balance the seesaw, the weight of each child needs to apply an equivalent force to the seesaw, but since they do not weigh the same, they need to sit at different lengths from the center. This can be set up as the following proportion:

$$W_{Chris} \times d_{Chris} = W_{Elliot} \times d_{Elliot}$$

The equation can be rearranged to find the relative distance Elliot must sit compared to Chris:

$$\frac{W_{Chris}}{W_{Elliot}} = \frac{d_{Elliot}}{d_{Chris}}$$

$$\frac{50 \text{ lb}}{75 \text{ lb}} = \frac{d_{Elliot}}{d_{Chris}} = \frac{2}{3}$$

Elliot must therefore sit two-thirds of the distance from the center than Chris is sitting. Choice *B* is the only choice which gives the appropriate ratio, 6 feet and 4 feet.

12. A: The system is in equilibrium, so neither weight will move. Both weights are pulling on the wheel and axle in different directions, so the equation can be used to find their equivalent force. Because both weights are pulling straight down from the wheel and axle, the force is perpendicular to lever arm, meaning $\sin(\theta) = 1$, simplifying the following equation for the wheel and axle's torque:

$$\tau_{total} = r_{axle} \times \sin(\theta_{axle}) \times F_{axle} - r_{wheel} \times \sin(\theta_{wheel}) \times F_{wheel}$$

$$\tau_{total} = \frac{1}{2} r_{wheel} \times (20 \text{ kg} \times g) - r_{wheel} \times (10 \text{ kg} \times g)$$

$$\tau_{total} = r_{wheel} \times 10 \text{ kg} \times g - r_{wheel} \times 10 \text{ kg} \times g = 0$$

Knowing the angles of the force, the relative forces, and the radius from the center allows predicting the behavior, even if exact values are unknown.

13. D: The baseball will have equal horizontal distance and displacement. Choice *A* is incorrect because the ball's path will be a downward parabolic curve due to gravity. To achieve equal horizontal and vertical distance, it would require an initial angle equating to a slope of 2. Choice *B* is incorrect because the ball will be under the effect of gravity during its flight, meaning that once it reaches half its total path, its height will not be less than the horizontal distance. So, once it falls back to its original height, the total vertical distance will also be less than the horizontal distance. Choice *C* is incorrect because its vertical displacement, once it falls back to the same height, is zero, while it will still have been displaced horizontally.

14. D: The kinetic energy of a pendulum at the bottom of its path is at its maximum. Because there are no losses to air resistance or friction, the total energy of the pendulum is simply the sum of kinetic and potential energy. Because the pendulum is at its lowest point, its potential energy is also at its minimum. Since no energy is being lost, the kinetic energy must therefore be at its maximum.

15. B: The sandbag can be suspended by 7.5 newtons of force. Because the system isn't ideal, it's necessary to include mechanical efficiency. Ideally, the pulley system would increase distance to reduce the force by a factor of three, but because the mechanical efficiency is also taking away from the input force, the actual mechanical advantage is only two-thirds of the ideal advantage, or an actual mechanical advantage of two.

Assembling Objects

1. C

2. D

3. C

4. B

5. D

6. B

7. A

8. B

9. D

10. D

11. A

12. B

13. A

14. C

15. B

Practice Test #2

General Science

1. What number on the pH scale indicates a neutral solution?
   a. 13
   b. 8
   c. 7
   d. 2

2. Which of the following compounds is an example of a strong base?
   a. HCl
   b. HNO3
   c. HBr
   d. NaOH

3. What is the primary difference between the Earth's inner core and outer core?
   a. The inner core is made of nickel, while the outer core is made of iron.
   b. The inner core is solid, while the outer core is molten.
   c. The inner core is inaccessible to people, while the outer core is accessible to people.
   d. The inner core is hot, while the outer core is cool.

4. The separation of a single radioactive element to create a powerful force is known as what type of process?
   a. Nuclear fission
   b. Nuclear meltdown
   c. Thermodynamics
   d. Proton pump inhibitor

5. Combining two or more atoms of a single radioactive element to create a powerful force is known as what type of process?
   a. Nuclear fission
   b. Nuclear meltdown
   c. Nuclear fusion
   d. Nuclear dynamic

6. In what type of chemical reaction is heat absorbed from the surrounding environment?
   a. Exothermic
   b. Endothermic
   c. Rusting
   d. Freezing

7. Which of the following is an example of friction?
   a. A rubber band that is pulled taut then snapped
   b. Jumping on the end of a diving board
   c. Throwing a baseball at an angle
   d. Applying a car's brakes at a red light

8. Isaac is reducing strawberry jam to make a dessert sauce. He takes the jam out of the refrigerator and spoons some into a saucepan. The jam is a cold, thick glob and takes a few minutes to slowly fall off the spoon and into the pan. Isaac turns the stove top on medium heat and the jam begins to turn runny and bubble. What factor of the jam did the heat affect?
   a. Its flavor
   b. Its mass
   c. Its viscosity
   d. Its caloric energy

9. Objects in which molecules are closest together can be described as which of the following in comparison to objects in which molecules are farther apart?
   a. High vibrational
   b. Dense
   c. Forceful
   d. Pressurized

10. The process in which a gas becomes a liquid is known as which of the following?
   a. Melting
   b. Sublimation
   c. Vaporization
   d. Condensation

11. Which nerve system allows the brain to regulate body functions such as heart rate and blood pressure?
   a. Central
   b. Somatic
   c. Autonomic
   d. Afferent

12. Which statement is true?
   a. Radioactive decay products are all unstable.
   b. Radioactive decay never occurs naturally.
   c. Radioactive decay is a natural process.
   d. Radioactive decay occurs in half of a sample.

13. Which type of macromolecule contains genetic information that can be passed to subsequent generations?
   a. Carbohydrates
   b. Lipids
   c. Proteins
   d. Nucleic acids

14. In which situation is an atom considered neutral?
   a. The number of protons and neutrons are equal.
   b. The number of neutrons and electrons are equal.
   c. The number of protons and electrons are equal.
   d. There are more electrons than protons.

15. Which type of matter has molecules that cannot move within its substance and breaks evenly across a plane caused by the symmetry of its molecular arrangement?
   a. Gas
   b. Crystalline solid
   c. Liquid
   d. Amorphous solid

Arithmetic reasoning

1. There are x trees along the running trail in the local park. The number of trees at the zoo is 34 more than 4 times the number of trees along the trail. Which expression represents the number of trees at the zoo?
   a. $34 > 4x$
   b. $34 - 4x$
   c. $34 + 4x$
   d. $4x > 34$

2. Keisha is planning a company picnic for the end of quarter celebration. She has a budget of $1,275 for renting a pavilion and purchasing catered food. There are 45 people in the office that will attend the picnic, and the cost of renting a pavilion is $310. Which of the following inequalities shows how to find the amount of money, x, that Annie can spend on a catered lunch for each employee?
   a. $45x + 310 \geq 1,275$
   b. $45x - 310 \leq 1,275$
   c. $45x - 310 \geq 1,275$
   d. $45x + 310 \leq 1,275$

3. Erin and Katie work at the same ice cream shop. Together, they always work less than 21 hours a week. In a week, if Katie worked two times as many hours as Erin, how many hours could Erin work?
   a. Less than 7 hours
   b. Less than or equal to 7 hours
   c. More than 7 hours
   d. Less than 8 hours

4. A study of adult drivers finds that it is likely that an adult driver wears his seatbelt. Which of the following could be the probability that an adult driver wears his seat belt?
   a. 0.90
   b. 0.05
   c. 0.25
   d. 0

5. The following set represents the test scores from a university class: {35, 79, 80, 87, 87, 90, 92, 95, 95, 98, 99}. If the outlier is removed from this set, which of the following is TRUE?
   a. The mean and the median will decrease
   b. The mean and the median will increase
   c. The mean and the mode will increase
   d. The mean and the mode will decrease

6. The mass of the Moon is about 7.348×10^{22} kilograms and the mass of Earth is 5.972×10^{24} kilograms. How many times greater is Earth's mass than the Moon's mass?

    a. 8.127×10^1

    b. 8.127

    c. 812.7

    d. 8.127×10^{-1}

7. What is the mode for the grades shown in the chart below?

Science Grades	
Jerry	65
Bill	95
Anna	80
Beth	95
Sara	85
Ben	72
Jordan	98

    a. 65

    b. 33

    c. 95

    d. 90

8. If a new monitor cost $300 with a 40% discount, what was its original price?

    a. $420

    b. $500

    c. $460

    d. $340

Use the following information to answer the next three questions, rounding to the closest minute. Eva Jane is practicing for an upcoming 5K run. She has recorded the following times (in minutes):

25, 18, 23, 28, 30, 22.5, 23, 33, 20

9. What is Eva Jane's approximate mean time rounded to the nearest minute?

    a. 26 minutes

    b. 19 minutes

    c. 25 minutes

    d. 23 minutes

10. What is the mode of Eva Jane's times?

    a. 16 minutes

    b. 20 minutes

    c. 23 minutes

    d. 33 minutes

11. What is Eva Jane's median time?
   a. 23 minutes
   b. 17 minutes
   c. 28 minutes
   d. 19 minutes

12. Which shapes could NOT be used to compose a hexagon?

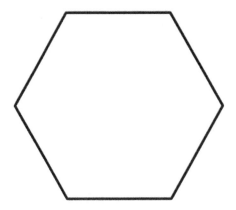

   a. Six triangles
   b. One rectangle and two triangles
   c. Two rectangles
   d. Two trapezoids

13. Which of the following expressions best exemplifies the distributive property?
   a. $6 + 8 + 5 = 6 + (8 + 5)$
   b. $4 \times 6 \times 3 = 3 \times 6 \times 4$
   c. $9 \times 1 = 9$
   d. $4 \times 8 + 3 = 4 \times 8 + 4 \times 3$

14. shoe salesclerk earns $8x + 0.00048y^2$ dollars, where x is the total number of hours worked and y represents her total sales in dollars. If she earned $200 dollars and had $500 in sales, how many hours did she work?
   a. 8
   b. 10
   c. 12
   d. 14

15. In a small university, the ratio of male to female students is 3:2. If there are 900 total students, how many are men?
   a. 540
   b. 360
   c. 750
   d. 150

Word Knowledge

1. *Conspicuous* most nearly means
    a. Scheme
    b. Obvious
    c. Secretive
    d. Ballistic

2. *Onerous* most nearly means
    a. Responsible
    b. Generous
    c. Hateful
    d. Burdensome

3. *Banal* most nearly means
    a. Inane
    b. Novel
    c. Painful
    d. Complimentary

4. *Contrite* most nearly means
    a. Tidy
    b. Unrealistic
    c. Contrived
    d. Remorseful

5. *Mollify* most nearly means
    a. Pacify
    b. Blend
    c. Negate
    d. Amass

6. *Capricious* most nearly means
    a. Skillful
    b. Sanguine
    c. Chaotic
    d. Fickle

7. *Paltry* most nearly means
    a. Appealing
    b. Worthy
    c. Trivial
    d. Fancy

8. *Shirk* most nearly means
    a. Counsel
    b. Evade
    c. Diminish
    d. Sharp

9. *Assuage* most nearly means
    a. Irritate
    b. Persuade
    c. Argue
    d. Soothe

10. *Tacit* most nearly means
    a. Unspoken
    b. Shortened
    c. Tenuous
    d. Regal

11. *Overt* most nearly means
    a. Obvious
    b. Concealed
    c. Central
    d. Annoying

12. *Composed* most nearly means
    a. Discarded
    b. Adjourned
    c. Misaligned
    d. Created

13. *Discern* most nearly means
    a. Disengage
    b. Distinguish
    c. Disembowel
    d. Dissociate

14. *Unrelated* most nearly means
    a. Pertinent
    b. Admissible
    c. Germaine
    d. Irrelevant

15. *Unassuming* most nearly means
    a. Ignorant
    b. Exposed
    c. Nondescript
    d. Blatant

Paragraph Comprehension

The next two questions are based on the following passage from Rhetoric and Poetry in the Renaissance: A Study of Rhetorical Terms in English Renaissance Literary Criticism *by D.L. Clark:*

> In both these respects the ancients felt that poetics, the theory of poetry, was different from rhetoric. As the critical theorists believed that the poets were inspired, they endeavored less to teach men to be poets than to point out the excellences which the

284

poets had attained. Although these critics generally, with the exceptions of Aristotle and Eratosthenes, believed the greatest value of poetry to be in the teaching of morality, no one of them endeavored to define poetry, as they did rhetoric, by its purpose. To Aristotle, and centuries later to Plutarch, the distinguishing mark of poetry was imitation. Not until the renaissance did critics define poetry as an art of imitation endeavoring to inculcate morality ...

1. What does the author say about one way in which the purpose of poetry changed for later philosophers?

   a.   The author says that at first, poetry was not defined by its purpose but was valued for its ability to be used to teach morality. Later, some philosophers would define poetry by its ability to instill morality. Finally, during the renaissance, poetry was believed to be an imitative art, but was not necessarily believed to instill morality in its readers.

   b.   The author says that the classical understanding of poetry dealt with its ability to be used to teach morality. Later, philosophers would define poetry by its ability to imitate life. Finally, during the renaissance, poetry was believed to be an imitative art that instilled morality in its readers.

   c.   The author says that at first, poetry was thought to be an imitation of reality, then later, philosophers valued poetry more for its ability to instill morality.

   d.   The author says that the classical understanding of poetry was that it dealt with the search for truth through its content; later, the purpose of poetry would be through its entertainment value.

2. The word *inculcate* in the second paragraph can be best interpreted as meaning which one of the following?

   a.   Imbibe
   b.   Instill
   c.   Implode
   d.   Inquire

3. The Roman Republic became the Roman Empire in 27 BCE when Augustus Caesar declared himself the first Emperor of Rome. Augustus expanded the territories controlled by Rome and initiated many reforms, which led to the Pax Romana, or Roman Peace. This period of peace and prosperity lasted for 200 years until the death of Marcus Aurelius, who anointed his incompetent son Commodus emperor in 176. Commodus was assassinated in 192, which set off a power play and a period of civil war. Government corruption became rampant and remained a problem until the Goths sacked Rome in 410.

According to the author, the event that directly led to civil war was:

   a.   Emperor Commodus' death
   b.   The death of Emperor Augustus
   c.   Widespread government corruption
   d.   Marcus Aurelius anointing Commodus emperor

4. Reverse osmosis water filtration is the process of forcing water through a very fine membrane to remove impurities. The process is used in both industrial and household settings. Reverse osmosis removes a wide variety of dissolved chemicals, including sodium, chloride, sulfate, and nitrate. These systems are very effective at removing bacteria and viruses.

Based on the passage, what is the *most accurate* statement?
a.   Water contains large amounts of bacteria.
b.   Reverse osmosis is a membrane filtration system.
c.   Dissolved chemicals must be removed from drinking water.
d.   Reverse osmosis systems employ advanced technology.

5. Ariana Grande's 2018 album *Sweetener* was a huge hit, debuting at number one on the U.S. Billboard 200. This was Grande's third U.S. #1 album after *Yours Truly* (2013) and *My Everything* (2014). Her 2020 album, *Positions*, became Ariana's fifth album to debut at number one. In 2021 it was announced that Grande would play Glinda in the film *Wicked*.

What year was *Sweetener* released?
a.   2014
b.   2021
c.   2018
d.   2013

6. She was so full of rage that no one dared to stand in her path. With fists bald and hair in a jumbled mess, she began to shout at any poor individual who happened to come to close. Everyone stared at her with bewilderment. No one seemed to know what this woman was raving about, but the saddest thing was, no one seemed to care to help.

What point of view is this passage told in?
a.   First-person
b.   Second-person
c.   Third-person
d.   Fourth-person

7. "Those boys do act beastly," stated the young girl very casually. For she had just seen, for the eighth time that day, three little boys, not more than eight-years-old each, hurl with great accuracy and surprising speed a volley of snowballs at a group of people walking through the park. They laughed hard as two dogs took off running away from them and sent the men in the group yelling to get them back.

What action caused the girl to think the boys behaved "beastly?"
a.   The fact that they threw snowballs at other people.
b.   The fact that they were laughing.
c.   The fact that the men were yelling.
d.   The fact that the boys were playing with snow

8. There once was a ship that went out to sea in an attempt to catch a great sea monster. There were over 150 men aboard this mighty vessel, but that would be a number far too few for what lay ahead.

What figure of speech is being used here?
    a. Hyperbole
    b. Metaphor
    c. Flashback
    d. Foreshadowing

9. This cat is green and that pear is red. Those oranges are not orange, but they should be! This baby is yellow, that kiwi is blue, and that pineapple is as a big as a tree. Everyone one of those clouds are gold while everyone of those bushed are solid black.

Which description does NOT fit with the others?
    a. The cat is green
    b. The pineapple tree is big
    c. The clouds are gold
    d. The oranges are not orange

10. The sky was sunny that day. There was not a rain cloud in the sky. Sweet birds could be heard chirping in every direction. Suddenly, without warning, there was a large gust of wind that bent the trees! The birds all fluttered away in an instant as if they knew what was coming. Large massive clouds seemed to appear out of nowhere and rain came down in sheets.

What is the tone of this passage?
    a. Excited
    b. Happy
    c. Anxious
    d. Nervous

Mathematics Knowledge

1. What is the volume of the cylinder below? Use 3.14 for π.

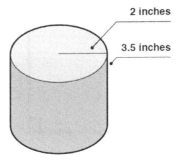

    a. 18.84 in^3
    b. 45.00 in^3
    c. 70.43 in^3
    d. 43.96 in^3

2. What is the solution to the following system of linear equations?

$$2x + y = 14$$

$$4x + 2y = -28$$

   a. $(14, -28)$
   b. $(2, -4)$
   c. All real numbers
   d. No solution

3. Which of the following is perpendicular to the line $4x + 7y = 23$?
   a. $y = -\frac{4}{7}x + 23$
   b. $y = \frac{7}{4}x - 12$
   c. $4x + 7y = 14$
   d. $y = -\frac{7}{4}x + 11$

4. What is the solution to the following system of equations?

$$2x - y = 6$$

$$y = 8x$$

   a. $(1, 8)$
   b. $(-1, 8)$
   c. $(-1, -8)$
   d. All real numbers

5. What is the missing length x?

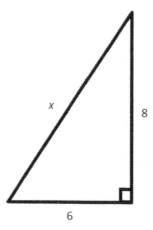

   a. 6
   b. −10
   c. 10
   d. 100

6. Which of the following is perpendicular to the line $5x + 8y = 17$?
   a. $y = 2x - 21$
   b. $y = -\frac{10}{3}x + 4$
   c. $x + 11y = \frac{1}{3}$
   d. $y = \frac{8}{5}x + 26$

7. What is the equation of the line that passes through the two points $(-3, 7)$ and $(-1, -5)$?
   a. $y = 6x + 11$
   b. $y = 6x$
   c. $y = -6x - 11$
   d. $y = -6x$

8. Simplify $(y^8)(y^{13})$.
   a. y^{104}
   b. y^{21}
   c. $2y^{21}$
   d. y^5

9. Simplify $\frac{z^{93}}{z^{31}}$.
   a. z^{62}
   b. z^3
   c. z^{31}
   d. z^{124}

10. What is the value of $6x^3 - 8y^3$ if $x = 1.25$ and $y = 2.1$?
    a. 62.36925
    b. −1,988.65725
    c. −4,139.757
    d. −62.36925

11. Solve the equation $5(x - 4) + 17 = 5x - 16$.
    a. No solution
    b. $x = 16$
    c. $x = -13$
    d. All real numbers

12. Simplify $(x^5)^{10}$.
    a. x^5
    b. x^{15}
    c. x^{50}
    d. x^2

13. What is the area of the following figure?

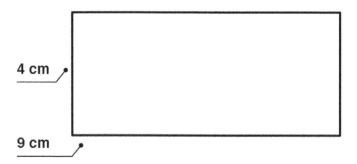

a. 26 cm
b. 36 cm
c. 13 cm^2
d. 36 cm^2

14. What is the volume of the given figure?

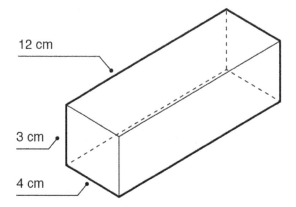

a. 36 cm^2
b. 144 cm^3
c. 72 cm^3
d. 144 cm^2

15. What type of units are used to describe surface area?
a. Square
b. Cubic
c. Single
d. Quartic

Electronics Information

1. For a capacitor in a series circuit, what happens to the current and voltage when the voltage across the capacitor increases?

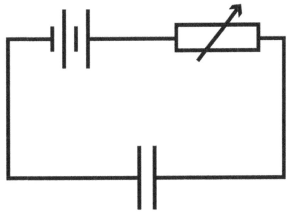

   a. The current across the capacitor decreases and the voltage decreases.
   b. The current across the capacitor remains constant and the voltage increases.
   c. The current across the capacitor increases and the voltage gradually increases.
   d. The current across the capacitor decreases and the voltage gradually increases.

2. A transformer consists of two or more inductors around a magnetic core. What is a transformer typically used for?
   a. It raises voltage or current
   b. It functions as a resistor
   c. It increases power output
   d. It stores energy

3. Which of the following statements about AC and DC circuits is correct?
   a. AC power supplies have zero average voltage.
   b. DC can be stepped up or down with a transformer.
   c. A DC power supply in series with a capacitor is continuous.
   d. An inductor coil stores up electrical charges from AC.

4. How often does the polarity of a 50 hertz AC current change?
   a. The polarity remains constant
   b. 25 times per second
   c. 50 times per second
   d. 100 times per second

5. Between a long, thin wire and a short, thick wire, which has a greater resistance?
   a. Not enough information to determine.
   b. The long, thin wire has a greater resistance.
   c. Both wires have the same resistance.
   d. The short, thick wire has a greater resistance.

6. What does a circuit breaker do?
   a. It acts as a central point for the circuit.
   b. It manages the current and resistances of the circuit.
   c. It recycles energy in the circuit.
   d. It shuts off the circuit in the event of a surge.

7. What units are capacitance measured?
   a. Farads
   b. Hertz
   c. Ohms
   d. Volts

8. What is the name of the process that adds impurities to semiconductors to alter their properties?
   a. Alteration
   b. Reconduction
   c. Doping
   d. Energizing

9. Which phrase describes the direction that current will flow through in this transistor?

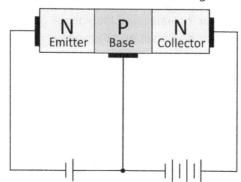

   a. Into the collector, through the base, out of the emitter
   b. Into the base, out of the emitter and the collector
   c. Into the collector, stored in the base until needed, out of the emitter
   d. Into the emitter, through the base, out of the collector

10. What device is this?

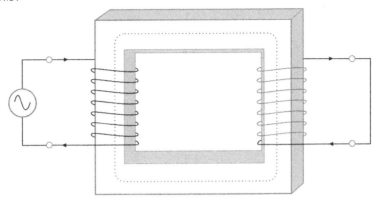

a. Motor
b. Inductor
c. Electromagnet
d. Transformer

11. What device is this?

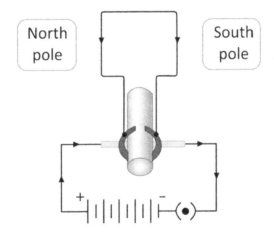

a. Generator
b. Motor
c. Capacitor
d. Rectifier

12. Which of these metals is the most effective conductor of electricity?
a. Aluminum
b. Steel
c. Copper
d. Brass

13. What is the current passing through the 3-ohm resistor?

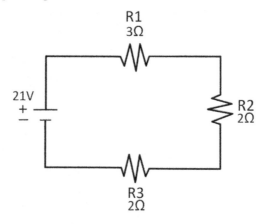

a. 3 amps
b. 7 amps
c. 5 amps
d. 10 amps

14. What is the current passing through the 2-ohm resistor on path B?

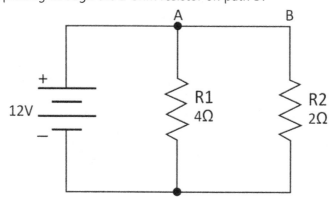

a. 6 amps
b. 2 amps
c. 3 amps
d. 1 amp

This content is provided exclusively for test preparation purposes and does not imply our support of any particular religious, political, or scientific point of view. Copyright © APEX Publishing. You have been licensed one copy of this document for personal use only. Any other reproduction or redistribution is strictly prohibited. All rights reserved.

15. What is the current passing through the 6-ohm resistor on path B if path C is disconnected?

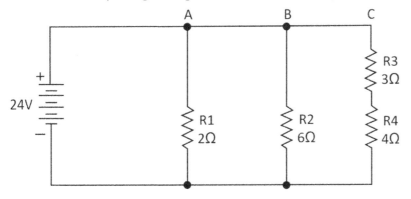

a. 2 amps
b. 4 amps
c. 3 amps
d. 12 amps

Auto Information

1. Which of the following tools is used to achieve a specific tightness for screws and bolts?
   a. Torque wrench
   b. Electric screwdriver
   c. Ratchet wrench
   d. Pliers

2. When jump-starting a dead vehicle battery, what is the correct order to connect the clamps?

   I. Dead battery positive terminal
   II. Unpainted metal on chassis away from dead battery
   III. Working battery positive terminal
   IV. Working battery negative terminal

   a. I to III, then II to IV
   b. I to III, then IV to II
   c. I to IV, then III to II
   d. I to IV, then II to IV

3. Which of the following materials provides the most resistance to vertical bending as a ceiling joist?
   a. Dimensional lumber
   b. I-joist
   c. Plywood joist
   d. Fiberboard

4. In the engine's cooling system, where is excess coolant stored when the vehicle is at rest?
   a. The radiator hoses
   b. The expansion tank
   c. The radiator
   d. The reservoir

5. Which of the following does NOT reduce the lifespan of a typical lead-acid battery?
   a. Cold weather
   b. Engine heat
   c. Hydrogen gas leak
   d. Water evaporation

6. What does the brake system's master cylinder do?
   a. Converts brake pedal depression to hydraulic pressure in brake lines
   b. Stores additional hydraulic fluid
   c. Increases braking power
   d. Transfers hydraulic fluid to the brakes

7. Which of the following primarily maintains the vehicle's ride height?
   a. Struts
   b. Shocks
   c. Springs
   d. Tires

8. Which engine stroke cycle is shown here?

   a. Exhaust stroke
   b. Power stroke
   c. Intake stroke
   d. Compression stroke

9. What is the ideal balance of antifreeze to water in the coolant's mixture?
   a. 40/60
   b. 25/75
   c. 50/50
   d. 70/30

10. How do car breaks work to stop a vehicle from moving?
    a. By turning off the axil
    b. By clamping the axil with break-pads
    c. By turning the wheels the opposite direction
    d. By squeezing the wheel drum with break-pads

Shop Information

1. What is a mechanical tool used to rapidly spin wood while a worker carves into it called?
    a. Rotary saw
    b. Chisel
    c. Drill press
    d. Lathe

2. Which type of tool would be used to make holes in sheet metal?
    a. Drill
    b. Die and punch
    c. Planer
    d. Hacksaw

3. Which type of electrical saw uses a reciprocating blade to cut curves in wood?
    a. Jig saw
    b. Table saw
    c. Chain saw
    d. Band saw

4. Which piece of personal protective equipment would be most useful when carrying several pieces of lumber?
    a. Apron
    b. Steel-toe boots
    c. Gloves
    d. Goggles

5. When using a fire extinguisher, where should the user aim the nozzle?
    a. Toward the middle of the fire
    b. Toward the top of the fire
    c. About 2 feet in front of the fire
    d. Toward the base of the fire

6. What would a bubble level with the following appearance indicate about the surface that it is on?

    a. The right side of the surface needs to be raised in order to be level.
    b. The left side of the surface needs to be raised in order to be level.
    c. The front side of the surface needs to be raised in order to be level.
    d. The surface is level.

7. Which material would be most appropriate for using a keyhole saw?
    a. Sheet metal
    b. Wooden logs
    c. Stone
    d. Drywall

8. What is a common use of a jackhammer?
    a. To fasten metal plates together
    b. To break apart concrete
    c. To move dirt like a shovel
    d. To make holes in metal plates

9. To protect the wearer from certain chemical spills, such as sulfuric acid, some aprons are made with which material?
    a. Cotton
    b. Denim
    c. PVC plastic
    d. Leather

10. Which part of this hammer is circled?

    a. Face
    b. Claw
    c. Cheek
    d. Throat

Mechanical Comprehension

1. A hammer used to drive a nail behaves like a third-class lever. The hammer's head weighs 1 kilogram (assuming the handle has no mass) and the length from the hammer to its fulcrum is 30 centimeters. If the hammer was pushed at a point 10 centimeters away from its fulcrum and generates a reactionary force of 100 N in the head, how much force is being applied?
   a. 33.33 N
   b. 100 N×m
   c. 300 N
   d. There is not enough information to determine the applied force.

2. What type of lever is a bench being lifted on one side?
   a. First-class
   b. Second-class
   c. Third-class
   d. Fourth-class

3. A small sailboat is being carried along by an eastern wind at a speed of 8 knots, while a water current is pulling it northwards at 6 knots. Assuming the wind and water current are completely efficient, what is the ship's total speed?
   a. 10 knots
   b. 12 knots
   c. 14 knots
   d. 15 knots

4. Small electric motors are used in many cooking appliances, such as mixers. One device has a motor that outputs 300 watts of mechanical power and produces 10 newton-meters of torque. What is the magnitude of the angular velocity from this motor?
   a. 15
   b. 30
   c. 150
   d. 3,000

5. Using a single pulley, a 10-kilogram weight is hoisted up 4 meters. How much energy is transferred to the weight, assuming no friction?
   a. 40 J
   b. 98.1 J
   c. 196.2 J
   d. 392.4 J

6. What is the weight of a person, standing on flat ground with no additional forces or movement, if the normal force between them and the ground is 880 newtons?
   a. 8,632.8 g
   b. 897.0 g
   c. 89.7 kg
   d. 8.97 kg

7. A child swings a toy airplane on a string in circles over their head. If they swing the toy faster, which of the following occurs?

    a.   The tension in the string increases.
    b.   The tension in the string remains the same.
    c.   The tension in the string decreases.
    d.   The tension is eliminated.

8. Which of the following symbols used in physics represents an angular quantity?

    a. μ
    b. v
    c. ω
    d. ρ

9. If two objects of different masses ($m_1 > m_2$), moving perpendicularly to each other at different velocities ($v_1 > v_2$), collide, which set of equations predicts the resulting magnitude of velocity and direction? Assume the objects behave as point masses in a completely inelastic collision, with no frictional forces.

    a. $v = \sqrt{\left(\dfrac{p_{1x}+p_{2x}}{m_1+m_2}\right)^2 + \left(\dfrac{p_{1y}+p_{2y}}{m_1+m_2}\right)^2}$ and $\theta = \tan^{-1}\dfrac{\left(\dfrac{p_{1y}+p_{2y}}{m_1+m_2}\right)}{\left(\dfrac{p_{1x}+p_{2x}}{m_1+m_2}\right)}$

    b. $v = \dfrac{p_{1x}+p_{2x}}{m_1+m_2} + \dfrac{p_{1y}+p_{2y}}{m_1+m_2}$ and $\theta = \tan\dfrac{\left(\dfrac{p_{1y}+p_{2y}}{m_1+m_2}\right)}{\left(\dfrac{p_{1x}+p_{2x}}{m_1+m_2}\right)}$

    c. $v = \dfrac{p_{1x}+p_{2x}}{m_1+m_2} + \dfrac{p_{1y}+p_{2y}}{m_1+m_2}$ and $\theta = \tan^{-1}\dfrac{\left(\dfrac{p_{1x}+p_{2x}}{m_1+m_2}\right)}{\left(\dfrac{p_{1y}+p_{2y}}{m_1+m_2}\right)}$

    d. $v = \sqrt{\left(\dfrac{p_{1x}+p_{2x}}{m_1+m_2}\right)^2 + \left(\dfrac{p_{1y}+p_{2y}}{m_1+m_2}\right)^2}$ and $\theta = \tan\dfrac{\left(\dfrac{p_{1x}+p_{2x}}{m_1+m_2}\right)}{\left(\dfrac{p_{1y}+p_{2y}}{m_1+m_2}\right)}$

10. A hydraulic car jack functions using pistons and an incompressible hydraulic fluid to increase input force, as shown below:

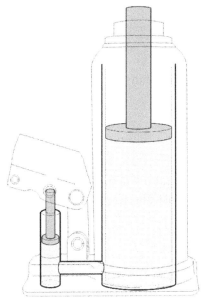

The area of the small piston head is 5 cm² and the large piston is 150 cm². If a small vehicle exerts a downwards force of 4,500 newtons on the large piston, what force must be applied to the small piston to lift it?
   a. 90 N
   b. 150 N
   c. 450 N
   d. 900 N

11. The MA of the lever of this system is 3. The MA of the pulley in this system is 4. What is the total MA of this compound machine
   a. 12
   b. 7
   c. 3/4
   d. 1

12. Which support column of this bridge is bearing the most weight?

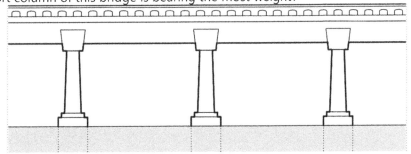

   a. The left column
   b. The central column
   c. The right column
   d. They are all equal

13. The quality of how well an object can bend without breaking is called what?
    a. Absorption
    b. Heat conduction
    c. Malleability
    d. Flexibility

14. When you squeeze a ball between your hands, what type of force are you applying?
    a. Compression
    b. Friction
    c. Tension
    d. Air pressure

15. You swing a sledgehammer at a wall and break through it. At the moment the sledgehammer impacts the wall, the wall itself exerts a force against the sledgehammer's impact. Which of Newton's laws of motion does this moment represent?
    a. First law
    b. Second law
    c. Third law
    d. Fourth law

Assembling Objects

For questions 1–7, which answer choice best shows how the objects on the left will connect if the letters for each object are put together?

1.

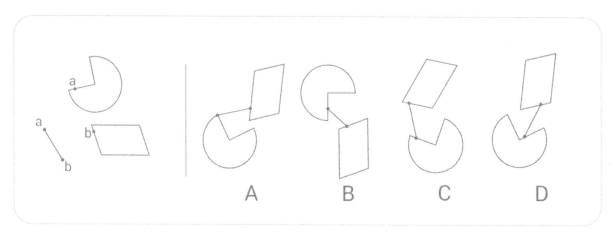

2.

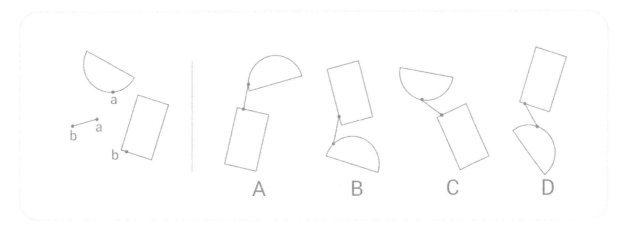

3.

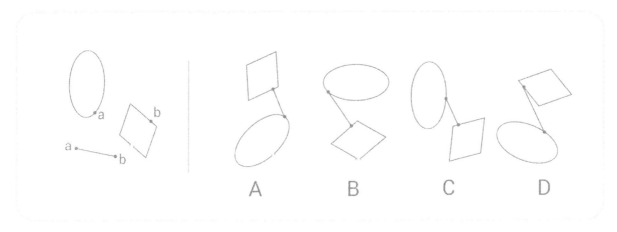

4.

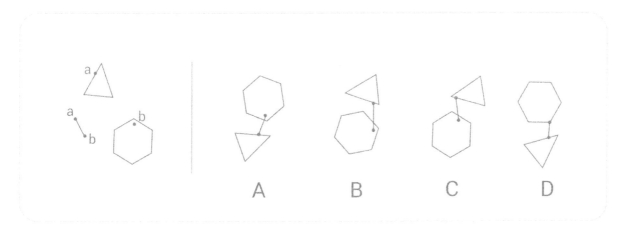

5.

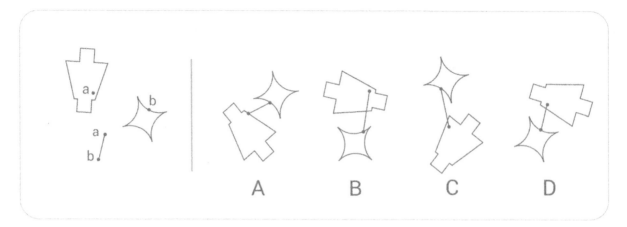

6.

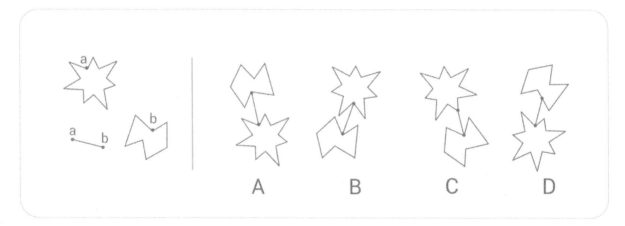

7.

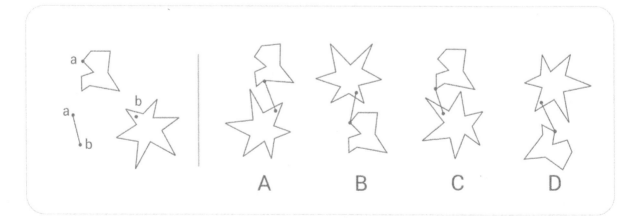

For questions 8–15, which answer choice shows the way the shapes on the left of the image would best fit together?

8.

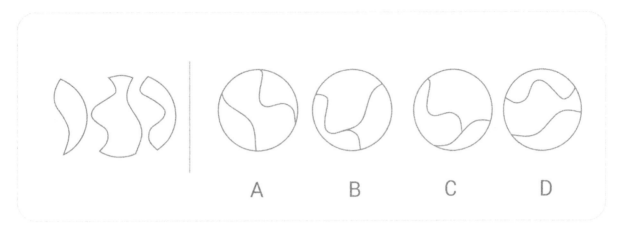

9.

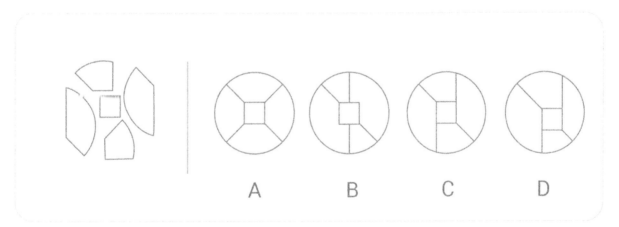

10.

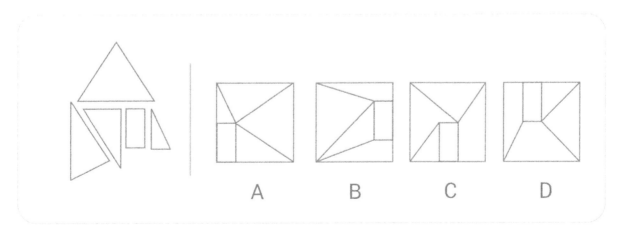

11.

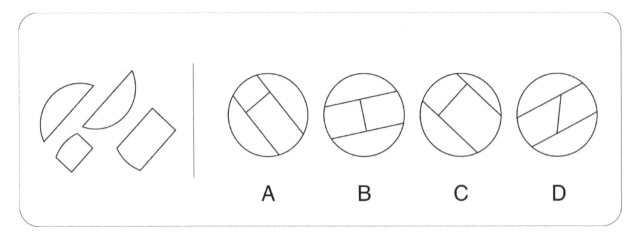

12.

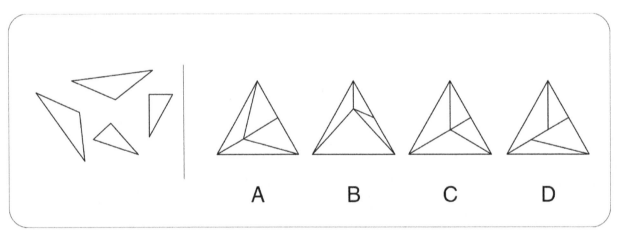

13.

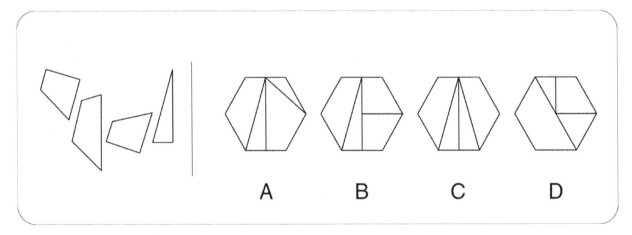

14.

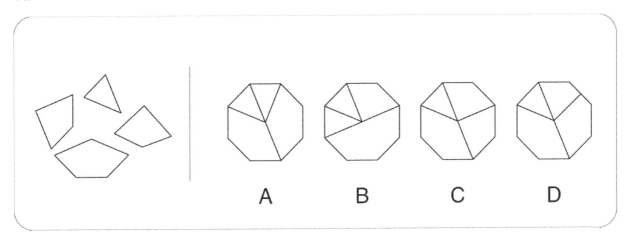

15.

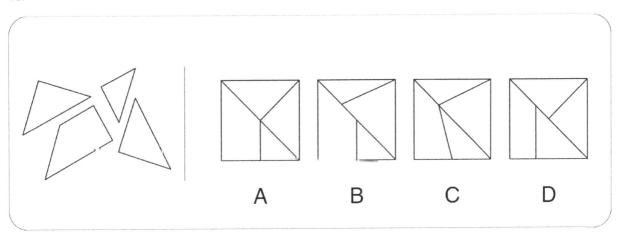

Answer Explanations #2

General Science

1. C: The pH scale goes from 0 to 14. A neutral solution falls right in the middle of the scale at 7. Choice *A*, a value of 13, indicates a strong base. Choice *B*, a value of 8, indicates a solution that is weakly basic. Choice *D*, a value of 2, indicates a strong acid.

2. D: A strong base dissociates completely and forms the anion OH^-. All bases include a hydroxide group (OH) in their formula, but not all basic compounds will dissociate completely; only strong bases dissociate completely into ions. Choices *A*, *B*, and *C* all represent strong acids and have a hydrogen atom, which is always present in acidic compounds.

3. B: The inner and outer core are both made up of the same elements (iron and nickel), extremely hot, and inaccessible to humans. However, the inner core is solid while the outer core is molten.

4. A: Nuclear fission occurs when the nucleus of a radioactive atomic element is split by force. This causes an explosive release of energy. A nuclear meltdown refers to any accident that occurs at a core reactor in a power plant; these are catastrophic to the surrounding geographic areas, as extreme amounts of radiation are released. Thermodynamics is the physical science of energetic relationships. Proton pump inhibitors are a class of drugs that limit intracellular pumping activity, typically to manage digestive issues.

5. C: Nuclear fusion occurs when lower-weight radioactive elements fuse and release energy rather than splitting, which occurs in the nuclear fission process. Nuclear meltdowns refer to catastrophic events at nuclear power plants, while nuclear dynamic is not a true physical science term.

6. B: Endothermic reactions consume heat from the surrounding environment to complete the chemical reaction and create a product. For example, holding an ice cube in one's hand and allowing it to melt into water (because of body heat) is an example of an endothermic reaction. Exothermic reactions release heat into the environment (i.e., igniting wood to create a fire). Rusting occurs when iron reacts to oxygen over a period of time. Freezing is a type of exothermic reaction.

7. D: Friction refers to any force that works opposite to an initial force, typically to slow down or stop the initial force. In this example, the brakes are working in the opposite direction from the way the wheels are rolling. Pulling a rubber band, jumping, and throwing a ball are adding more force to an event, rather than working in the opposite manner.

8. C: Viscosity describes a liquid material as being thick because of internal friction or resistance. Heating a viscous substance makes it less viscous. In this case, the jam was viscous when it was cold and in a more solid state; it resisted moving off the spoon and onto the pan. As Isaac heated the jam, it became less viscous and was able to easily move around the pan. Low temperature is unlikely to quickly change the taste, caloric content, or flavor of a food item.

9. B: An object's density refers to how close its molecules physically are to one another in space. Typically, as objects cool and solidify, density increases and volume decreases. As objects heat up and become more liquid or gaseous, volume increases and density decreases.

10. D: Condensation refers to the process of a gas becoming a liquid. For example, surface water on blades of grass evaporates when ground and surface temperatures are warmer in the daytime; as nighttime temperatures cool the blades of grass, the water vapor condenses into droplets. Melting is the process of turning a solid into a liquid. Vaporization is the process of turning a liquid into a gas. Sublimation is the process of turning a solid into a gas.

11. C: The autonomic nerve system is part of the efferent division of the peripheral nervous system (PNS). It controls involuntary muscles, such as smooth muscle and cardiac muscle, which are responsible for regulating heart rate, blood pressure, and body temperature. The central nervous system, Choice *A*, includes only the brain and the spinal cord. The somatic nervous system, Choice *B*, is also part of the efferent division of the PNS and controls voluntary skeletal muscle contractions, allowing for body movements. The afferent division of the PNS, Choice *D*, relays sensory information within the body.

12. C: As stated in the example, radioactive decay is a natural process that happens spontaneously when atoms of one element decay into another element. Choice *A* is incorrect because decay occurs until the atoms in the sample reach a stable state. It is not limited to only half of the atoms; therefore, Choice *D* is incorrect.

13. D: Nucleic acids include DNA and RNA, which are strands of nucleotides that contain genetic information. Carbohydrates, Choice *A*, are made up of sugars that provide energy to the body. Lipids, Choice *B*, are hydrocarbon chains that make up fats. Proteins, Choice *C*, are made up of amino acids that help with many functions for maintaining life.

14. C: Atoms are considered neutral when the number of protons and electrons is equal. Protons carry a positive charge, and electrons carry a negative charge. When they are equal in number, their charges cancel out, and the atom is neutral. If there are more electrons than protons, or vice versa, the atom has an electric charge and is termed an ion. Neutrons do not have a charge and do not affect the electric charge of an atom.

15. B: Solids have molecules that are packed together tightly and cannot move within their substance. Crystalline solids have atoms or molecules that are arranged symmetrically, making all of the bonds of even strength. When they are broken, they break along a plane of molecules, creating a straight edge. Amorphous solids do not have the symmetrical makeup of crystalline solids, so they do not break evenly. Gases and liquids both have molecules that move around freely.

Arithmetic Reasoning

1. C: The phrase "more than" represents addition in mathematics. Four times the number of trees along the trail is equal to $4x$. Therefore, the expression representing the number of trees at the zoo is $34 + 4x$.

2. D: If the cost of the lunch is x, then, because there will be 45 employees attending the lunch, the cost of catering will be $45x$. The sum of this amount and $310, which is the cost of the pavilion, has to be less than or equal to the budgeted amount of $1,275. Therefore, the correct inequality is:

$$45x + 310 \leq 1,275$$

3. A: Let x be the unknown, the number of hours Erin can work. We know Katie works $2x$, and the sum of all hours is less than 21. Therefore, $x + 2x < 21$, which simplifies into $3x < 21$. Solving this results in the inequality $x < 7$ after dividing both sides by 3. Therefore, Erin can work less than 7 hours.

4. A: The probability of 0.9 is closer to 1 than any of the other answers. The closer a probability is to 1, the greater the likelihood that the event will occur. The probability of 0.05 shows that it is very unlikely that an adult driver will wear their seatbelt because it is close to zero. A zero probability means that it will not occur. The probability of 0.25 is closer to zero than to one, so it shows that it is unlikely an adult will wear their seatbelt.

5. B: The outlier is 35. When a small outlier is removed from a data set, the mean and the median increase. The first step in this process is to identify the outlier, which is the number that lies away from the given set. Once the outlier is identified, the mean and median can be recalculated. The mean will be affected because it averages all of the numbers. The median will be affected because it finds the middle number, which is subject to change because a number is lost. The mode will most likely not change because it is the number that occurs the most, which will not be the outlier if there is only one outlier.

6. A: Division can be used to solve this problem. The division necessary is:

$$\frac{5.972 \times 10^{24}}{7.348 \times 10^{22}}$$

To compute this division, divide the constants first then use algebraic laws of exponents to divide the exponential expression.

This results in about 0.8127×10^2, which written in scientific notation is 8.127×10^1.

7. C: The mode for a set of data is the value that occurs the most. The grade that appears the most is 95. It's the only value that repeats in the set.

8. B: If the discount is 40%, the price of the monitor must be multiplied times 1.40 to account for the discount. Since subtracting 40% from 100% is 60%, and 60% is $\frac{3}{5}$ of 100%, this can be restated as multiplying $300 by $\frac{5}{3}$:

9. C: The mean is found by adding all the times together and dividing by the number of times recorded:

$$25 + 18 + 23 + 28 + 30 + 22.5 + 23 + 33 + 20 = 222.5$$

Divide by 9 to get 24.7. Rounding to the nearest minute, the mean is 25.

10. C: The mode is the time from the data set that occurs most often. The number 23 occurs twice in the data set, while all others occur only once, so the mode is 23.

11. A: To find the median of a data set, you must first list the numbers from smallest to largest, and then find the number in the middle. If there are two numbers in the middle, add the two numbers together and divide by 2. Putting this list in order from smallest to greatest yields 18, 20, 22.5, 23, 23, 25, 28, 30, and 33, where 23 is the middle number, so 23 minutes is the median.

12. C: A hexagon can be formed by any combination of the given shapes except for two rectangles. There are no two rectangles that can make up a hexagon.

13. D: The distributive property is used when a number is being multiplied by a term or terms inside of parentheses. The number outside the parentheses is multiplied, or distributed, to each term, and the results are added or subtracted as called for.

14. B: First, plug 500 in for y, which represents her sales. Then, set this expression equal to 200, her total earnings:

$$8x + 0.00048(500)^2 = 200$$

Simplify:

$$8x + 120 = 200$$

Solving this for x gives $x = 10$, and therefore, she worked 10 hours.

15. A: Let m represent the number of male students and f represent the number of female students. Therefore,

$$m + f = 900$$

Also, we know that $\frac{m}{f} = \frac{3}{2}$ because the ratio of male to female students is 3/2. Cross-multiplying this proportion results in $2m = 3f$, or $f = \frac{2}{3}f$. Substituting this into the first equation results in $m + \frac{2}{3}m = 900$, so $\frac{5}{3}m = 900$. Solving this for m results in $m = 540$ students.

Word Knowledge

1. B: *Conspicuous* means to be visually or mentally obvious. Something conspicuous stands out, is clearly visible, or may attract attention.

2. D: *Onerous* most closely means burdensome or troublesome. It usually is used to describe a task or obligation that may impose a hardship or burden, often which may be perceived to outweigh its benefits.

3: A: Something *banal* lacks originality and may be boring and trite. For example, a banal compliment is likely to be a platitude. Like something that is inane, a banal compliment might be meaningless and lack a convincing quality or significance.

4. D: *Contrite* means to feel or express remorse, or to be regretful and interested in repenting. The noun *contrition* refers to severe remorse or penitence.

5. A: *Mollify* means to soothe, pacify, or appease. It usually is used to refer to reducing the anger, or softening the feelings or temper, of another person. For example, a customer service associate may need to mollify an irate customer who is furious about the defect in their purchase.

6. D: *Capricious* mean to be erratic or unpredictable in behavior, which is similar to fickle. *Fickle* means to change spontaneously or act erratically.

7. C: *Paltry* means something is trivial or insignificant. It is often used as an adjective to describe a very small or meager amount (of money, in particular).

8. B: *Shirk* means to evade or avoid and is often used in the context of shirking a responsibility, duty, or work.

9. D: *Assuage* means to soothe or comfort, as in to assuage one's fears. It can also mean to lessen or make less severe, or to relieve. For example, an ice pack on a swollen knee may assuage the pain.

10. A: *Tacit* means unspoken or implied. Tacit approval, for example, occurs when agreement or approval is understood without explicitly stating it.

11. A: *Overt* can mean "obvious," "open," or "apparent." For example: The woman's anger was very *overt* when she began yelling into her phone.

12. D: *Compose* means to create or write. *Create* and *compose* are similar because both involve the making or remaking of something new or different.

13. B: *Discern* means to distinguish or determine based on what can be seen. Someone can *discern* that it is going to rain when they see dark clouds in the sky.

14. D: *Unrelated* means that a thing does not pertain or relate to another thing. A teacher might tell a student talking about football during a science lecture that that is an *unrelated* topic.

15. C: *Unassuming* means something lacks distinctive or noteworthy features. A leaf of green grass in a green field would be *unassuming*.

Paragraph Comprehension

1. B: The author says that the classical understanding of poetry dealt with its ability to be used to teach morality. Later, philosophers would define poetry by its ability to imitate life. Finally, during the renaissance, poetry was believed to be an imitative art that instilled morality in its readers. The rest of the answer choices improperly interpret this explanation in the passage. Poetry was never mentioned for use in entertainment, which makes Choices *D* and *E* incorrect. Choices *A* and *C* are incorrect because they mix up the chronological order.

2. B: The correct answer choice is Choice *B*, *instill*. Choice *A*, *imbibe*, means to drink heavily, so this choice is incorrect. Choice *C*, *implode*, means to collapse inward, which does not make sense in this context. Choice *D*, *inquire*, means to investigate. This option is better than the other options, but it is not as accurate as *instill*. Choice *E*, *idolize*, means to admire, which does not make sense in this context.

3. A: A power vacuum, an empty seat in the government, was created when Emperor Commodus died. This led directly to civil war as people raced to fill this vacancy. Choice *B* cannot be correct because Emperor Augustus's death occurred long before the civil war. Choice *C* is incorrect because government corruption became an issue after the war started. Marcus Aurelius anointing Commodus as emperor did not lead to the civil war, although Commodus is said to have been an incompetent emperor.

4. B: The passage stated that reverse osmosis is a membrane filtration system, or Choice *B*. The passage did not mention the number of bacteria, choice *A*, or the need to remove chemicals, choice *C*. Finally, although it might be implied, the technology is not described as being advanced, Choice *D*.

5. C: The passage stated that *Sweetener* was released in 2018, or choice *C*. Choice *A*, 2014, was the year of the release of *My Everything*, while choice *B*, 2021, was when Ariana was announced to play Glinda in the film *Wicked*. Choice *D*, 2013, was when *Yours Truly* was released.

6. C: This story gives several clues to indicate that the point of view is from the third person perspective. The use of the pronouns "she" and "her" indicate that it is not told from the woman's point of view. Since the passage refers to everyone around that also indicates that this is told in the third person. Also, the absence of the pronouns "I", "you", "we", etc. shows that this was not written in the first- or second-person point of view. Fourth-person point of view is not a real term used in English writing.

7. A: The key to correctly answering this question is found by looking at the word "for." This indicates why the girl said that the boys were acting beastly. It is clear that it was not due to the boys merely playing with snow, but the fact that they were throwing it at people who were out enjoying the day. Laughter, Choice *B*, was a result of their beastly behavior, but not the reason why they were described that way. The men yelling, Choice *C*, has nothing to do with the girl's thoughts toward the boys.

8. D: Choice *A*, hyperbole, is an exaggeration that is used to emphasize a point and not to be taken literal. Choice *B*, metaphor, is a type of speech used to compare one thing with another. Choice *C*, flashback, is the opposite of what this passage uses because a flashback is when something done in the past is brought to the readers attention. Choice *D*, foreshadowing, is right because it leads the reader to think that something bad will happen to this crew of people trying to catch a monster. It does not tell of their peril, but it is hinted at.

9. B: The answer is Choice *B* because every other thing described is described based upon what color it is. The pineapple tree being as big as a tree has nothing to do with its color.

10. A: This passage begins as though it could be happy, but the majority of the text is about the terrible storm that appeared. Choice *B*, excited, is correct because of the words that are used to describe the scene: "Suddenly," "without warning," and "in an instant" all indicate a rapid change and suspense. Choices *C* and *D* are incorrect because there is nothing in the passage about anxiety or feelings of nervousness.

Mathematics Knowledge

1. D: The volume for a cylinder is found by using the formula:

$$V = \pi r^2 h = (3.14)(2 \text{ in})^2 \times 3.5 \text{ in} = 43.96 \text{ in}^3$$

2. D: This system can be solved using the method of substitution. Solving the first equation for y results in $y = 14 - 2x$. Plugging this into the second equation gives $4x + 2(14 - 2x) = -28$, which simplifies to $28 = -28$, an untrue statement. Therefore, this system has no solution because no x-value will satisfy the system.

3. B: The slopes of perpendicular lines are negative reciprocals, meaning their product is equal to –1. The slope of the line given needs to be found. Its equivalent form in slope-intercept form is $y = -\frac{4}{7}x + \frac{23}{7}$, so its slope is $-\frac{4}{7}$. The negative reciprocal of this number is $\frac{7}{4}$. The only line in the options given with this same slope is:

$$y = \frac{7}{4}x - 12$$

4. C: This system can be solved using substitution. Plug the second equation in for y in the first equation to obtain $2x - 8x = 6$, which simplifies to $-6x = 6$. Divide both sides by 6 to get $x = -1$, which is then substituted back into either original equation to obtain $y = -8$.

5. C: The Pythagorean theorem can be used to find the missing length x because it is a right triangle. The theorem states that $6^2 + 8^2 = x^2$, which simplifies into $100 = x^2$. Taking the positive square root of both sides results in the missing value $x = 10$.

6. D: The slopes of perpendicular lines are negative reciprocals, meaning their product is equal to –1. The slope of the line given needs to be found. Its equivalent form in slope-intercept form is $y = -\frac{5}{8}x + \frac{17}{8}$, so its slope is $-\frac{5}{8}$. The negative reciprocal of this number is $\frac{8}{5}$. The only line in the options given with this same slope is $y = \frac{8}{5}x + 26$.

7. C: First, the slope of the line must be found. This is equal to the change in y over the change in x, given the two points. Therefore, the slope is -6. The slope and one of the points are then plugged into the point-slope form of a line:

$$y - y_1 = m(x - x_1)$$

This results in:

$$y - 7 = -6(x + 3)$$

The –6 is distributed and the equation is solved for y to obtain:

$$y = -6x - 11$$

8. B: To combine exponential expressions through multiplication with the same base, add the exponents and keep the same base. Therefore, the answer is:

$$y^{13+8} = y^{21}$$

9. A: To divide exponential expressions with the same base, subtract the exponents and keep the same base. Therefore, the answer is:

$$z^{93-31} = z^{62}$$

10. D: To evaluate the expression, substitute 1.25 in for x and 2.1 in for y. Therefore, $6(1.25)^3 - 8(2.1)^3$ needs to be evaluated. Following order of operations, the exponents need to be evaluated first. Therefore, the expression is equal to:

$$6(1.953125) - 8(9.261)$$

Next, evaluate the multiplication, obtaining:

$$11.71875 - 74.088 = -62.36925$$

11. A: To solve the equation, first distribute the 5 to remove the parentheses:

$$5x - 20 + 17 = 5x - 16$$

Collecting like terms results in:

$$5x - 3 = 5x - 16$$

Subtracting $5x$ from both sides results in an untrue equation: $-3 = -16$. Therefore, this equation is never true. Hence, there is no solution.

12. C: An exponential expression raised to another power is simplified by multiplying the exponents. Therefore,

$$(x^5)^{10} = x^{50}$$

13. D: The area for a rectangle is found by multiplying the length by the width. The area is also measured in square units, so the correct answer is Choice *D*. The number 26 in Choice *A* is incorrect because it is the perimeter. Choice *B* is incorrect because the answer must be in centimeters squared. The number 13 in Choice *C* is incorrect because it is the sum of the two dimensions rather than the product of them.

14. B: The volume of a rectangular prism is found by multiplying the length by the width by the height. This formula yields an answer of 144 cubic centimeters. The answer must be in cubic units because volume involves all three dimensions. Each of the other answers have only two dimensions that are multiplied and one dimension is forgotten, as in Choice *A*, where 12 and 3 are multiplied, or have incorrect units, as in Choice *D*.

15. A: Surface area is a type of area, which means it is measured in square units. Cubic units are used to describe volume, which has three dimensions multiplied by one another. Quartic units describe measurements multiplied in four dimensions.

Electronics Information

1. D: The current decreases while the voltage increases. A capacitor resists changes to its voltage by inducing a current against the existing current to counter the change in voltage. Because voltage and current are directly proportional, when the voltage in the circuit is increased, the capacitor will create a current against the circuit's flow, effectively decreasing current, to counteract the increased voltage. This continues until the capacitor's plates reach maximum charge, at which point current through the capacitor drops to zero.

2. A: A transformer functions on the principle of electromagnetic induction. In a transformer with two inductive coils, current flows through one coil, creating a magnetic flux through the transformer's core, which induces a current in the *other* inductor's coil, proportional to the winding ratio, or turns between the two coils. Choice *B* is incorrect; although the conductive elements of a real transformer possess resistance, a transformer serves to increase or decrease voltage (by conversely decreasing or increasing current). A transformer does not increase power. An increase of voltage through a transformer is accompanied by a proportional decrease in current, so there is no change in power, making Choice *C* incorrect. Choice *D* is also incorrect; although energy moves into the transformer core via magnetic fields, it is transmitted to the secondary coil as electric current again, so the transformer is not storing up energy.

3. A: The average voltage of an AC power supply is zero. This is because the voltage changes in polarity periodically, so an AC source with a peak voltage of 170 volts will be nearly +170 volts at one moment, but −170 volts at the next, resulting in zero average voltage. Because electromotive force is only created in the presence of changing magnetic flux, a transformer does not work on direct current, which creates a constant magnetic flux. Choice *C* is incorrect because a capacitor works by specifically creating a small gap between two conductive surfaces to allow charges to build up on each side. Choice *D* is incorrect because an inductor doesn't store up charges—the coil builds a magnetic field.

4. D: The polarity of the 50 hertz AC current changes 100 times per second. The frequency of an AC current represents the number of cycles per second. The polarity has to change twice to go back to its original state—representing one cycle—so the rate it changes polarity is 100 times per second.

5. B: The long, thin wire has a greater resistance. For any real wire, there is some amount of resistance. Because resistance in a conductor is proportional to the length of a wire and inversely proportional to its cross-sectional area, if a wire's diameter decreases, there is a greater resistance. Although the length and cross-sectional area are unknown, the resistance of the longer, thinner wire must be greater than the short, thick wire, so Choice *B* is correct.

6. D: A circuit breaker's function is to detect a surge in current and shut off the circuit to prevent excessive damage. It does not act as a central point in the circuit or recycle energy, so Choices *A* and *C* are incorrect. Choice *B* is incorrect because while circuit breakers may be able to monitor the current passing through them to detect a surge, they do not manage the current or resistances of the circuit.

7. A: Capacitance is measured in farads. Hertz measures frequency, ohms measure resistance, and volts measure voltage, so Choices *B, C,* and *D* are all incorrect.

8. C: Adding impurities to a semiconductor to alter its properties is called doping.

9. D: In this transistor, current would enter into the emitter, then be passed through the base into the collector and continue along the circuit.

10. D: This device is a transformer, used to convert high voltage to low voltage or the other way around. Inductors and electromagnets are present as components but are not the whole device, so Choices *B* and *C* are incorrect. Choice *A*, an electric motor, is a different device.

11. B: This device is an electric motor for converting electrical energy into mechanical energy. Generators convert mechanical energy into electrical energy, but a motor converts electrical energy into mechanical energy, so Choice *A* is incorrect. A capacitor temporarily stores electricity, so Choice *C* is incorrect. A rectifier is a type of circuit for converting AC signal to DC, so Choice *D* is incorrect.

12. C: Out of these metals, copper is the most efficient conductor of electricity. Aluminum is not quite as effective as copper, so Choice *A* is incorrect. Brass contains copper, but other metals mixed into it reduce its conductivity, making Choice *D* incorrect as well. Steel is less effective than copper or aluminum, so Choice *B* is also incorrect.

13. A: In a single series circuit like this, the current is affected by all three resistors additively. Add together the resistances in ohms of all three resistors: $3 + 2 + 2 = 7$ ohms. Then divide the voltage by the total resistance: $\frac{21}{7} = 3$ amps.

14. B: On a parallel circuit, the equivalent resistance is still equal to all resistors across all paths added together. Add together the resistances in both resistors: $4 + 2 = 6$ ohms. Then divide the voltage by the total resistance: $\frac{12}{6} = 2$ amps.

15. C: This is a parallel circuit, but if path C is disconnected, then its resistors are ignored when calculating the equivalent resistance. Add together the resistances from paths A and B: $2 + 6 = 8$ ohms. Then divide the voltage by the total resistance: $\frac{24}{8} = 3$ amps.

Auto Information

1. A: Torque wrenches give feedback on the amount of force used, indicating the tightness of a fastener. None of the other tools necessarily tell the user the torque applied. A torque wrench may be a ratcheting type or non-ratcheting type, but a ratcheting wrench itself is not necessarily a torque wrench. Electric screwdrivers can drive screws or bolts (depending on its attachment), while pliers are used to hold or twist smaller parts.

2. B: The correct order is to clamp the positive terminal of the dead battery to the positive terminal of the working battery, then the negative terminal of the working battery to ground on the chassis for the dead vehicle. The last connection completes the circuit; by connecting it to the dead vehicle's chassis instead of directly to the negative terminal of the dead battery, it minimizes the explosion risk of igniting flammable gas from the battery. The vehicle's chassis carries current to the battery by the grounding strap.

3. B: Wooden I-beams called I-joists are a type of engineered wood that's lighter and stiffer than traditional lumber, making it ideal for ceiling joists. Plywood and fiberboard are also engineered woods; however, they do not provide the same strength as I-beams.

4. D: The coolant reservoir stores the excess coolant to allow the engine to warm up to optimal temperatures and then releases coolant as needed. The expansion tank provides a space for coolant to expand when it heats up. The radiator and hoses do not store extra coolant.

5. A: Cold weather can reduce the effectiveness of a battery, but it does not directly cause damage to the battery components. Engine batteries can lose effectiveness by a loss of water due to evaporation, reducing electrolyte volumes in the cells, and hydrogen gas leaking.

6. A: The master cylinder converts force from the brake pedal into hydraulic pressure in the brake lines. The reservoir is what stores additional hydraulic fluid. The vacuum booster is what increases or amplifies the force from the pedal to the master cylinder. The brake lines carry hydraulic fluid, transferring force from the master cylinder to the brakes.

7. C: The vehicle's springs are a primary determinant of its ride height and allow the vehicle to ride over small bumps and disturbances and return to its normal height. The struts support the vehicle's spring and are a structural component of the vehicle, mounted directly to the vehicle's chassis. Shocks instead dampen motion and reduce the vehicle's bouncing.

8. B: The diagram shows the power stroke step of the engine cycle. While the intake stroke looks similar, this diagram does not show the intake valve opening, so Choice *C* is incorrect.

9. C: A 50/50 ratio of antifreeze to water is the best at ensuring the coolant has sufficient anti-freeze and anti-corrosion capabilities.

10. D: A car move when the axil of the car works to turn the wheels. The way car breaks get the axil to stop turning the wheels is by clamping break-pads onto the wheel drums. This creates friction and forces the wheels to come to a stop. Turning the car off would not stop the wheels from moving. Nothing clamps onto the axil and the wheel drum provides much more surface area to push against. Also, there is no safe or effective way to turn a car wheel the opposite direction while moving.

Shop Information

1. D: A lathe is used to spin wood rapidly while a worker carves designs into it. Choice *A* is incorrect because a rotary saw makes cuts into wood using a rapidly spinning blade. Choice *B* is incorrect because a chisel is used to cut into wood by using some type of forceful impact, although chisels can also be used with lathes to cut into spinning wood. Choice *C* is incorrect because a drill press is used to make hole in wood, like any other type of drill.

2. B: A die and punch are used to make holes in thin pieces of metal, such as sheet metal, by placing the metal on the die and the punch on top of the metal above the die and hammering it through. Choices *A* and *C* are incorrect because a drill and a planer should generally be used in woodwork, instead of metal work. Choice *D* is incorrect because a hacksaw would be used to cut length away from metal piping, not to make holes in it.

3. A: A jig saw uses a reciprocating blade (one that goes back and forth) to make irregular cuts, such as curves, in wood. Choices *B*, *C*, and *D* are incorrect because table saws, chain saws, and band saws all use spinning or rotary motion, instead of reciprocating motion, to cut wood.

4. C: Gloves would be the most useful personal protective equipment when carrying lumber, as they can protect the hands from splinters, cuts, and painful pressure. Choice *A* is incorrect because aprons, while somewhat helpful in reducing abrasions, are primarily used to protect the body from chemicals. Choice *B* is incorrect because steel-toe boots would be useful only if you dropped the lumber on your foot. Choice *D* is incorrect because goggles would provide no meaningful protection in carrying lumber.

5. D: When using a fire extinguisher, the user should aim toward the base of the fire, because the surface and substances at the base are the materials that are actually on fire. Choices *A*, *B*, and *C* are incorrect because aiming at these areas may cause the user to miss the area that is burning and thus prove ineffective.

6. A: A bubble level with a bubble that is left of center means that the right side needs to be raised in order to be level. Choice *B* is incorrect because raising the left side would cause the bubble to move farther to the left. Choice *C* is incorrect because the left-right orientation of the bubble is not related to the height of the front side of a surface. Choice *D* is incorrect because the bubble would need to be in the center to indicate that the surface is level.

7. D: A keyhole saw is a handheld saw with a small blade, designed for cutting holes in soft woods or in drywall. Choices *A*, *B*, and *C* are incorrect because sheet metal, wooden logs, and stone would be either too hard or too large for a keyhole saw to cut through.

8. B: A jackhammer is commonly used to break apart concrete. Choice *A* is incorrect because fastening metal plates is a common use of rivets and a rivet gun. Choice *C* is incorrect because jackhammers are not used to pick up or move materials. Choice *D* is incorrect because making holes in metal plates is a common use for a die and punch.

9. C: Aprons made with PVC plastic are resistant to many acids, including hydrochloric and sulfuric acids. Choices *A*, *B*, and *D* are incorrect because neither cotton, nor denim, nor leather would provide necessary protection from these types of chemicals.

10. B: The circled area of this hammer is called the claw. Choice *A* is incorrect because the face is the flat area of the hammer used to strike surfaces. Choice *C* is incorrect because the cheek is the side area of the hammer's head. Choice *D* is incorrect because the throat is the area of the hammer's head that sits between the cheek and the face.

Mechanical Comprehension

1. C: The force applied to the handle is 300 N. Recall the equation for the mechanical advantage of levers:

$$F_E a = F_L b$$

If the force applied 10 centimeters from the fulcrum results in a 100-newton force in the head, which is 30 centimeters from the fulcrum, the effort can be solved for by rearranging the equation and substituting in the known values:

$$F_E = \frac{F_L b}{a} = \frac{100 \text{ N} \times 30 \text{ cm}}{10 \text{ cm}}$$

$$F_E = 100 \text{ N} \times 3 = 300 \text{ N}$$

Therefore, the unknown force must be 300 newtons.

2. B: A bench being lifted on one side is an example of a second-class lever, because the load (the weight of the bench) is between the fulcrum (the legs of the bench still touching the ground) and the effort (the end being grabbed). Choice *D* is incorrect because there is no fourth class of levers, while Choice *A* and *C* are incorrect because the effort is neither across from the load nor between the fulcrum and the load.

3. A: To find the ship's total speed, the northwards and eastwards components must both be considered. To find the sum of these components, Pythagorean's theorem can be used, taking the square root of the sum of the squared components:

$$a^2 + b^2 = c^2$$

$$(8 \text{ kn})^2 + (6 \text{ kn})^2 = c^2$$

$$c = \sqrt{(8 \text{ kn})^2 + (6 \text{ kn})^2}$$

$$c = \sqrt{64 \text{ kn}^2 + 36 \text{ kn}^2} = \sqrt{100 \text{ kn}^2} = 10 \text{ kn}$$

Since the question is asking for the boat's speed rather than velocity, the answer is 10 knots.

4. B: The magnitude of maximum angular velocity of the motor is 30. This can be found by using the relationship of power (P), torque (τ), and angular velocity (ω):

$$P = \tau \times \omega$$

Since the power and torque are known values, they can be plugged in to the equation to solve for ω:

$$\omega = \frac{P}{\tau} = \frac{300 \text{ W}}{10 \text{ N} \times \text{m}} = \frac{300 \frac{\text{kg} \times \text{m}^2}{\text{s}^3}}{10 \frac{\text{kg} \times \text{m}}{\text{s}^2} \times \text{m}} = \frac{300 \frac{\text{kg} \times \text{m}^2}{\text{s}^3}}{10 \frac{\text{kg} \times \text{m}^2}{\text{s}^2}} = 30 \text{ s}^{-1}$$

Therefore, the magnitude of the angular velocity is 30.

5. D: 40 joules of work are performed in raising the weight. To solve this question, first recall that work is a measurement of energy, or the force applied along a distance. The equation for work is:

$$W = F \times d$$

Solving for work yields:

$$W = \left(m_{weigh} \times g\right) \times 4 \text{ m}$$

$$W = \left(10 \text{ kg} \times 9.81 \frac{\text{m}}{\text{s}^2}\right) \times 4 \text{ m}$$

$$W = 98.1 \text{ N} \times 4 \text{ m} = 392.4 \text{ J}$$

Therefore, the work done is 392.4 joules

6. C: The weight of the person would be 89.7 kg. Because they are standing still on flat ground, the normal force acting against them will be equal to their mass times acceleration due to gravity:

$$F_N = m \times g$$

$$880 \text{ N} = m \times 9.81 \frac{\text{m}}{\text{s}^2}$$

$$m = \frac{880 \text{ N}}{9.81 \text{ m/s}^2} = 89.7 \text{ kg}$$

7. A: The tension in the string increases as the toy is swung faster. The tension in the string is proportional to the forces pulling against it. When the toy is swung faster, the pulling force from the toy increases because the rate of change in the linear velocity of the toy increases—in other words, its linear acceleration increases, increasing its force.

8. C: The symbol ω, lowercase omega, represents angular velocity, the rate at which an object rotates relative to a fixed point. Choice A, μ or lowercase mu, represents the coefficient of friction. Choice B, v, represents linear velocity. Choice D, ρ or lowercase rho, is used to represent density.

9. A: Recall the equation for momentum is $p = mv$, and the subscripts x and y indicate the separate perpendicular components. Finding the magnitude of the sum velocity using x and y components is analogous to finding the hypotenuse of a right triangle. By separating the x and y components of two vectors, they can be used in the Pythagorean theorem to find the magnitude of the vector sum. Similarly, finding the resulting angle uses the inverse of tangent. The equations for the magnitude and direction of velocity are therefore:

$$v = \sqrt{\left(\frac{p_{1x} + p_{2x}}{m_1 + m_2}\right)^2 + \left(\frac{p_{1y} + p_{2y}}{m_1 + m_2}\right)^2}$$

$$\theta = \tan^{-1} \frac{\left(\dfrac{p_{1y} + p_{2y}}{m_1 + m_2}\right)}{\left(\dfrac{p_{1x} + p_{2x}}{m_1 + m_2}\right)}$$

10. B: To lift the vehicle, 150 newtons of force are required. Because the piston being pressed is much smaller while the fluid volume is static, it has to be pressed farther to displace the larger piston, so the force applied against the vehicle will be greater than the input. The hydraulic pistons multiply force equal to the ratio of the cross-sectional area of the pistons. The large piston's cross-sectional area (150 cm²) is

320

30 times that of the small piston (5 cm²), so the input force needs to be at least one-thirtieth of 4,500 newtons:

$$F_I = 4{,}500 \text{ N} \times \frac{5 \text{ cm}^2}{150 \text{ cm}^2} = \frac{4{,}500 \text{ N}}{30} = 150 \text{ N}$$

11. A: In compound machines, the total MA is the MA of all simple machine components multiplied together. $3 \times 4 = 12$ MA for the whole system.

12. D: Since all support structures are equidistant from each other, they are all supporting the same amount of weight.

13. D: Flexibility is how well an object can bend without breaking. Choice *C*, malleability, is similar, but refers to how well an object can be reshaped or reformed. Choice *A* refers to how much fluid an object can soak in. Choice *B* is how easily an object can heat up and transfer heat.

14. A: When you push an object inward from two opposite sides, that force is called compression. Friction is a resisting force when two surfaces touch, so Choice *B* is incorrect. Tension is a pulling force, so Choice *C* is incorrect. Air pressure is also a type of pushing force but does not necessarily push from two opposite sides, so Choice *D* is incorrect.

15. C: Newton's third law of motion says that for every action there is an equal and opposite reaction. In the instant the sledgehammer connects with the wall, the wall pushes back with an equal and opposite force to the impact.

Assembling Objects

1. C

2. D

3. B

4. A

5. D

6. A

7. B

8. D

9. C

10. A

11. A

12. C

13. B

14. C

15. A

Practice Test #3

Practice Test #3 is exclusively online. To access it, type in the link below or scan the QR code:

apexprep.com/bonus/asvab

The first time you access the page, you will need to register as a "new user" and verify your email address. If you encounter any problems, please email info@apexprep.com.

Index

Absolute Value, 53, 61, 62

Acceleration, 26, 216, 217, 262

Acid, 12, 19, 21, 45, 194, 197, 200, 207, 233, 241, 300, 302, 312

Activation Energy, 17, 197

Acute, 146, 147, 159, 162

Addition Principle of Equality, 112, 113

Addition Property of 0, 61

Addition Property of Equality, 104

Additive Inverse, 113

Adjacent Side, 159, 160

Aerobic Respiration, 45, 46

Air Compressor, 194

Air Resistance, 29, 32, 33, 219, 220, 221, 254, 255, 276, 279

Algebraic Equation, 107

Algebraic Expression, 56, 107, 108, 109, 113, 115

Alleles, 48, 49

Alpha Decay, 22

Alpha Particle, 22

Alternating Current (AC), 186

Alternative Hypothesis, 143

Amp, 175, 185, 298

Ampere (A), 175

Amphoteric, 21

Anaerobic Respiration, 45, 46

Angle-Angle-Angle (AAA), 162

Angle-Angle-Side (AAS), 162

Angle-Side-Angle (ASA), 162

Angular Momentum, 30, 43

Angular Momentum Quantum Number, 43

Angular Motion, 222

Angular Velocity, 222, 275, 303, 324, 325

Anion, 21, 43, 207, 312

Area, 78, 85, 151, 157, 166, 168

Arithmetic Average, 140

Asexual Reproduction, 47

Atom, 12, 15, 22, 39, 43, 44, 46, 175, 191, 282, 312, 313

Atomic Mass, 15

Atomic Mass Unit (amu), 20

Atomic Number, 12, 14, 15, 22, 43

Atoms, 43, 313

Automotive Lights, 200

Average, 10, 15, 26, 27, 51, 72, 75, 115, 132, 133, 135, 140, 141, 143, 196, 217, 226, 236, 264, 295, 320

Axes, 124, 125, 127, 216

Axis, 30, 31, 118, 119, 121, 124, 125, 126, 127, 129, 133, 134, 136, 137, 138, 168, 209, 222, 224, 262

Axle, 199, 200, 204, 223, 255, 279

Bar Graph, 126, 129, 133, 136, 137

Base, 10, 21, 40, 53, 54, 61, 75, 76, 103, 104, 120, 139, 149, 153, 158, 165, 170, 193, 281, 296, 301, 312, 319, 320, 323

Base-10 Number System, 53

Basic Shapes, 148

Battery, 40, 42, 177, 179, 180, 181, 184, 188, 194, 195, 196, 197, 200, 205, 250, 273, 274, 299, 300, 321, 322

Bell-Shaped, 134

Beta Radiation, 22

Bias, 143

Biases, 96

Bimodal, 132, 134

Brake Fluid, 198

Brake Fluid Leak, 203

Braking Systems, 198

Capacitance, 186, 187, 191, 250, 273, 296

Car Battery, 197, 207

Car Jack, 193, 305

Cartesian Coordinate System, 216

Catalysts, 17

Cation, 21, 43

Cause, 24, 34, 38, 40, 46, 47, 49, 91, 135, 194, 227, 242, 263, 322, 323

Cells, 44

Cellular Reproduction, 47, 263

Celsius (°C) Scale, 10

Census, 143

Centrifugal Force, 31, 262

Centripetal Force, 31, 262

Chemical Equations, 17

Chemical Properties, 12

Chemical Reaction, 12, 15, 16, 17, 22, 40, 175, 188, 197, 281, 312

Chlorophyll, 46

Chord, 158, 159

Y-Component, 28, 118

Greetings!

First, we would like to give a huge "thank you" for choosing us and this study guide for ASVAB exam. We hope that it will lead you to success on this exam and for your years to come.

Our team has tried to make your preparations as thorough as possible by covering all of the topics you should be expected to know. In addition, our writers attempted to create practice questions identical to what you will see on the day of your actual test. We have also included many test-taking strategies to help you learn the material, maintain the knowledge, and take the test with confidence.

We strive for excellence in our products, and if you have any comments or concerns over the quality of something in this study guide, please send us an email so that we may improve.

As you continue forward in life, we would like to remain alongside you with other books and study guides in our library. We are continually producing and updating study guides in several different subjects. If you are looking for something in particular, all of our products are available on Amazon. You may also send us an email!

Sincerely,
APEX Test Prep
info@apexprep.com

FREE

Free Study Tips Videos/DVD

In addition to this guide, we have created a FREE set of videos with helpful study tips. **These FREE videos provide you with top-notch tips to conquer your exam and reach your goals.**

Our simple request is that you give us feedback about the book in exchange for these strategy-packed videos. We would love to hear what you thought about the book, whether positive, negative, or neutral. It is our #1 goal to provide you with quality products and customer service.

To receive your **FREE Study Tips Videos**, scan the QR code or email freevideos@apexprep.com. Please put "FREE Videos" in the subject line and include the following in the email:

> a. The title of the book
>
> b. Your rating of the book on a scale of 1-5, with 5 being the highest score
>
> c. Any thoughts or feedback about the book

Thank you!

Made in the USA
Las Vegas, NV
11 December 2023

82502294R00188